Cereals and Pseudocereals in the 21st Century - Challenges and Opportunities

Edited by Eva Ivanišová

Published in London, United Kingdom

Cereals and Pseudocereals in the 21st Century - Challenges and Opportunities
http://dx.doi.org/10.5772/intechopen.1009073
Edited by Eva Ivanišová

Contributors
Ajeet Singh, Akash Fatima, Akhtar Hameed, Anand Kumar, Ayşe Usta, Beatrix Namulo Alweendo, Eva Ivanišová, Fawad Zafar Ahmad Khan, Felicitas Mantjodi Fwanyanga, Jana Žiarovská, Justina Itaveleni Shimwafeni, Kadadanamari S. Shashidhar, Kuume Benjamen Pendapala Enguwa, Laxmidas Verma, Lydia Ndinelao Horn, Mahipal Choubey, Maria Kapembe, Muhammad Ali Raza, Pramod Kumar Sharma, Ravindra Kumar, S. Bhuvaneshwari, Sagar, Sajad Hussain, Sehrish Ijaz, Silvia Farkasová, Ummara Waheed Khan, Zareen Sarfraz

First published in London, United Kingdom, 2026 by IntechOpen
IntechOpen is the global imprint of INTECHOPEN LIMITED, registered in England and Wales, registration number: 11086078, 167-169 Great Portland Street, London, W1W 5PF, United Kingdom

For EU product safety concerns: IN TECH d.o.o., Prolaz Marije Krucifikse Kozulić 3, 51000 Rijeka, Croatia, info@intechopen.com or visit our website at intechopen.com.

British Library Cataloguing-in-Publication Data
A catalogue record for this book is available from the British Library

Cereals and Pseudocereals in the 21st Century - Challenges and Opportunities
Edited by Eva Ivanišová
p. cm.

This title is part of the Agricultural Sciences Book Series, Volume 40
Topic: Agronomy and Horticulture
Series Editor: W. James Grichar
Topic Editors: Murat Helvaci and Olga Panfilova

Print ISBN 978-1-83635-698-1
Online ISBN 978-1-83635-697-4
eBook (PDF) ISBN 978-1-83635-699-8
ISSN 3029-052X

IntechOpen Book Series

Agricultural Sciences

Volume 40

Aims and Scope of the Series

The importance of agriculture cannot be overstated. It helps sustain life, as it gives us the food we need to survive and provides opportunities for economic well-being. Agriculture helps people prosper around the world and combines the creativity, imagination, and skill involved in planting crops and raising animals with modern production methods and new technologies. This series includes two main topics: Agronomy and Horticulture, and Animal Farming. This series will help readers better understand the intricacies of production agriculture and provide the new knowledge that is required to be successful. The success of a farmer in modern agriculture requires knowledge of events happening locally as well as globally that impact input decisions and ultimately determine net profit.

Meet the Series Editor

W. James Grichar has been employed with Texas A&M AgriLife Research for over 45 years with an emphasis on research in agronomy, plant pathology, and weed science. He obtained his BS from Texas A&M in 1972 and his Masters of Plant Protection in 1975. He has published 195 journal articles, over 330 research reports and briefs, 11 book chapters, and over 300 abstracts of profession meetings. He also directs research in many crops including corn, grain sorghum, peanuts, and sesame. He has held various positions in different professional societies including the American Peanut Research and Education Society, Southern Weed Science Society, and Texas Plant Protection Conference in addition to being Associate Editor for Peanut Science and Weed Technology. Significant accomplishments have included spearheading efforts to determine the optimum planting time for soybean production along the upper Texas Gulf Coast. These efforts have shown growers that soybean yields can be improved by 10 to 20% by following a late March to early April plant date. He also has been instrumental in developing a herbicide program for peanut production in the south Texas growing region. Through the development and use of herbicides that are effective against major weed problems in the south Texas region, peanut yields have increased by 25 to 30%.

Meet the Volume Editor

Doc. Ing. Eva Ivanišová, Ph.D., is an associate professor at the Institute of Food Sciences, Faculty of Biotechnology and Food Sciences, Slovak University of Agriculture in Nitra. She is also the scientific guarantor of the Food Incubator in the Research Centrum AgroBioTech, the Slovak University of Agriculture in Nitra. Her field of study is food technology, with a focus on plant raw materials and foodstuffs, foods with added value (enriched with medicinal plants and pseudocereals), lesser-known fruits and vegetables, and alternative food sources such as edible insects and flowers. Dr. Ivanišová has published more than 600 scientific works in international and domestic journals.

Contents

Preface

Cereals and pseudocereals are one of the most important groups of plants used in human nutrition. They have shaped the food systems of civilizations for millennia and still play a fundamental role in ensuring global food security. However, their importance extends beyond the traditional understanding as a source of energy; they are the subject of intensive scientific research in nutrition, food processing, agronomy, genetics, and molecular biology.

From a botanical point of view, cereals belong to the grass family (*Poaceae*), which includes wheat, barley, rye, oats, rice, and maize. Pseudocereals, such as quinoa (*Chenopodium quinoa*), amaranth (*Amaranthus spp.*), and buckwheat (*Fagopyrum esculentum*), do not belong to this family. However, their seeds are used similarly and are high in nutritional value. The different genetic origins of cereals and pseudocereals are reflected not only in their morphological features but also in the composition of proteins, starch content, micronutrients, and bioactive substances.

The genetic makeup of these crops plays a key role in shaping their qualitative and functional properties. The variability of cereal genomes, often characterized by polyploidy (for example, wheat), has enabled the breeding of varieties with higher yields, better technological quality, and increased resistance to biotic and abiotic stresses. In the case of pseudocereals, genetic diversity is an important source of traits such as higher levels of essential amino acids, gluten-free character, or increased tolerance to extreme growing conditions.

This book aims to provide a comprehensive view of cereals and pseudocereals, covering their biological, genetic, and nutritional properties, technological processing, and importance for human health. The publication integrates knowledge from classical breeding with modern genetic and genomic approaches. It highlights the potential of these crops to create sustainable food systems and rational nutrition in the context of a changing global environment.

Eva Ivanišová
Institute of Food Sciences,
Slovak University of Agriculture in Nitra,
Nitra, Slovak Republic

Section 1

Genome Editing and Functional Genomics in Cereals and Pseudocereals

Chapter 1

CRISPR and Gene Editing in Cereals and Pseudocereals: Current Progress and Future Directions

Akash Fatima, Zareen Sarfraz, Muhammad Ali Raza, Sajad Hussain, Akhtar Hameed, Fawad Zafar Ahmad Khan, Sehrish Ijaz and Ummara Waheed Khan

Abstract

CRISPR/Cas-based gene editing has rapidly revolutionized plant breeding, offering precise, efficient, and versatile tools to enhance traits in staple and emerging crops. This chapter explores the application of CRISPR and related gene-editing technologies in cereals and pseudocereals, focusing on advancements in improving yield, nutritional quality, biotic and abiotic stress tolerance, and grain pigmentation. The chapter begins with an overview of CRISPR/Cas systems, including Cas9, Cas12, and base editing tools, tailored for plant applications, especially during 2018–2025. It highlights successful case studies in major cereals such as rice, maize, and wheat, where gene editing has enabled the modification of key agronomic traits. Parallel progress in pseudocereals such as quinoa, buckwheat, and amaranth, though still emerging, is discussed in light of recent breakthroughs and bottlenecks, such as transformation efficiency and regulatory hurdles. Emphasis is placed on the integration of CRISPR with omics technologies and AI-assisted gene discovery to accelerate functional genomics and breeding pipelines. The chapter also considers ethical, regulatory, and public acceptance aspects of gene-edited crops, with special attention to policies in South Asia and developing economies. Finally, future directions are proposed to maximize the potential of CRISPR in underutilized pseudocereals, contributing to food security and climate-resilient agriculture.

Keywords: colored grains, anthocyanins, pseudocereals, functional foods, breeding, genomics, CRISPR, market potential

1. Introduction

Cereals and pseudocereals are vital to global food security, providing the majority of calories and essential nutrients to the world's population. Major cereals such as rice (*Oryza sativa* L.), wheat (*Triticum aestivum* L.), and maize (*Zea mays* L.) dominate the agricultural landscape, while pseudocereals such as quinoa

(*Chenopodium quinoa* Willd.), buckwheat (*Fagopyrum esculentum* Moench), and amaranth (*Amaranthus hypochondriacus* L.) have gained attention due to their superior nutritional profiles, gluten-free nature, and adaptability to marginal environments.

However, traditional breeding approaches often face limitations in improving complex traits such as drought tolerance, disease resistance, and grain quality due to polygenic inheritance, linkage drag, and long breeding cycles. The advent of genome-editing technologies, particularly CRISPR (Clustered Regularly Interspaced Short Palindromic Repeats)-Cas systems, has revolutionized crop improvement by enabling precise and efficient modification of target genes [1, 2].

CRISPR offers numerous advantages over earlier tools such as zinc finger nucleases (ZFNs) and transcription activator-like effector nucleases (TALENs) due to its simplicity, cost-effectiveness, and scalability. It allows scientists to knock out undesirable genes, fine-tune gene expression, and even correct alleles without introducing foreign DNA [3].

Over 250 CRISPR-edited plant species have now been reported globally, with cereals accounting for approximately 60% of applications [4]. This rapid expansion underscores the transformative potential of CRISPR in achieving food and nutritional security.

In this chapter, we explore the remarkable potential of CRISPR/Cas systems to address current and future challenges in cereal and pseudocereal breeding. We examine the molecular basis of CRISPR editing, showcase success stories in major crops, analyze the progress and prospects for pseudocereals, and assess integration with other cutting-edge technologies, such as omics and artificial intelligence.

2. CRISPR/Cas systems in plants

2.1 Historical background and evolution

The CRISPR/Cas system originated from prokaryotic immune defense mechanisms. In 2012, Jennifer Doudna and Emmanuelle Charpentier demonstrated the programmable nature of the CRISPR/Cas9 system, which marked the beginning of its rapid adoption in plant and animal biotechnology [5]. Since then, CRISPR has evolved into a family of tools beyond Cas9, including Cas12a (Cpf1), Cas13, and engineered variants for base and prime editing.

2.2 Mechanism of action

At its core, CRISPR/Cas functions as a molecular scissor guided by an RNA molecule. The single-guide RNA (sgRNA) directs the Cas enzyme to a specific genomic sequence adjacent to a protospacer adjacent motif (PAM). Upon recognition, the enzyme introduces a double-stranded break (DSB), which is then repaired by the plant's natural repair pathways, non-homologous end joining (NHEJ) or homology-directed repair (HDR). NHEJ often leads to gene knockouts via insertions or deletions, while HDR can be harnessed for gene replacement or insertion [4], as illustrated in **Figure 1**.

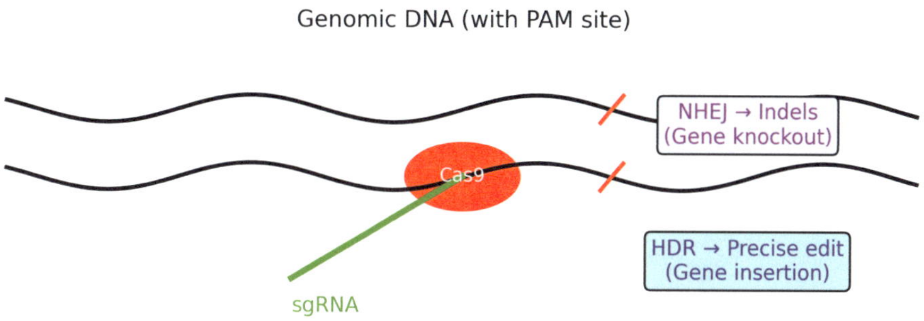

Figure 1.
Overview of the CRISPR/Cas mechanism in plants. The Cas nuclease, guided by a single-guide RNA (sgRNA), is directed to a specific genomic DNA sequence adjacent to a protospacer adjacent motif (PAM) site. It creates a double-strand break (DSB), which is then repaired by the plant's cellular repair pathways. Non-homologous end joining (NHEJ) often results in insertions or deletions (indels), leading to gene knockouts. Homology-directed repair (HDR) can be used to incorporate a donor DNA template for precise gene insertion or correction.

2.3 Variants of CRISPR/Cas tools for plants

CRISPR technology has rapidly expanded from the original Cas9 nuclease to a versatile toolbox optimized for multiple genome-editing outcomes in plants. Cas9 remains the most widely adopted nuclease across cereals due to its broad protospacer adjacent motif (PAM) compatibility and consistently high editing efficiency [5]. Cas12a, previously known as Cpf1, offers unique advantages, including the generation of staggered double-strand breaks and the ability to target AT-rich genomic regions – features that make it particularly useful for monocots such as *Oryza sativa* L. and *Zea mays* L. [6]. Beyond classical nucleases, base editors have enabled single-nucleotide conversions without inducing double-strand breaks, dramatically decreasing unwanted indels and allowing the correction of point mutations in regulatory and coding regions [7]. Prime editing further advances precision by introducing a Cas9-nickase fused to a reverse transcriptase, supporting targeted insertions and deletions without donor templates and expanding the scope of programmable genome modifications in cereals and pseudocereals [6, 7]. These editing modalities collectively allow fine-tuning of complex agronomic traits, moving crop improvement beyond knockout mutations toward rational genome design.

2.4 Delivery methods

Different CRISPR delivery methods have been developed to introduce editing reagents into plant cells, and the choice of method largely depends on crop species, laboratory resources, regulatory considerations, and whether DNA-free editing is desired. Agrobacterium tumefaciens-mediated transformation remains the most widely used approach because of its high success rate, relatively low cost, and stable transgene integration across cereals, including *Oryza sativa* L., *Triticum aestivum* L., and *Zea mays* L. Biolistic transformation (particle bombardment) is particularly useful for genotypes that are recalcitrant to Agrobacterium, as it enables direct delivery of plasmid DNA or ribonucleoprotein complexes into embryogenic calli, although this method may lead to fragmented or multiple insertions. Viral vectors are emerging as a promising transient delivery platform for CRISPR reagents, enabling rapid editing without stable genomic integration and, in certain cases, eliminating the need for tissue culture. DNA-free genome editing using ribonucleoproteins (RNPs) is gaining scientific and regulatory interest because it avoids the integration of foreign DNA and increases the chances of obtaining transgene-free edited plants, which are more likely to meet regulatory approval. According to Woo [8], RNP-mediated editing has already demonstrated high efficiency and a reduced off-target footprint. Although delivery efficiency still varies across cereals and pseudocereals, advances in tissue culture responsiveness, haploid induction editing, and meristem-targeted transformation continue to accelerate the practical deployment of CRISPR systems across difficult-to-edit crop species.

3. Applications in major cereals

Cereals such as rice (*Oryza sativa* L.), wheat (*Triticum aestivum* L.), and maize (*Zea mays* L.) form the cornerstone of global food systems. Their genetic improvement has been a central focus of CRISPR/Cas technology since its adoption in plant science. Gene editing has enabled researchers to modify key agronomic traits with precision, including yield, stress tolerance, nutrient content, and disease resistance. Below is an overview of CRISPR applications in each of these staple cereals.

3.1 Rice

Rice has served as a model crop for CRISPR implementation due to its small genome size, high transformation efficiency, and comprehensive genetic resources. One of the early breakthroughs involved the knockout of the *OsBADH2* gene, leading to enhanced fragrance through the accumulation of 2-acetyl-1-pyrroline (2AP), a key aroma compound [2], while editing *OsERF922* enhanced blast resistance by 40% under field trials [9]. Recent multiplex editing of *Gn1a*, *GS3*, and *GW2* resulted in 12–18% yield increases and recent studies have demonstrated multiplex CRISPR editing of yield-related genes in rice [10].

CRISPR has also been deployed to improve disease resistance in rice. The *OsERF922* gene, a negative regulator of blast resistance, was successfully edited, resulting in enhanced resistance to *Magnaporthe oryzae* L. without yield penalty [9]. Yield-related genes such as *Gn1a*, *GS3*, and *GW2* have been targeted in multiplex CRISPR constructs, showing additive effects on panicle size, grain length, and weight [4].

Another milestone was the development of semi-dwarf rice by disrupting *SD1*, mimicking the Green Revolution phenotype through targeted mutagenesis [11]. Additionally, nutrient fortification has been achieved by editing genes involved in phytic acid metabolism to enhance iron and zinc bioavailability [12].

3.2 Wheat

CRISPR editing in wheat is more complex due to its hexaploid genome, but several advances have been made. Targeted mutagenesis of all three homeologs of the *TaMLO* gene conferred resistance to powdery mildew (*Blumeria graminis* L.) in bread wheat [9]. This was a significant breakthrough in demonstrating multiplex editing in polyploid species.

Drought and salinity tolerance have been improved by editing *TaERF3* and *TaDREB2*, transcription factors involved in abiotic stress pathways [8]. The *Q* gene, a domestication-related gene affecting spike morphology and threshability, has also been edited to produce compact spikes with higher grain yield [13].

Gene editing is being explored to reduce immunogenic gluten peptides. Disruption of α-gliadin genes using CRISPR/Cas9 has led to wheat lines with significantly reduced immunoreactivity, holding promise for gluten-sensitive individuals [14]. Editing *TaNAC2* enhanced drought tolerance, maintaining 15% higher grain weight under stress CRISPR/Cas9 has been used to target multiple stress-related loci in wheat, demonstrating improved drought tolerance in edited lines [15].

3.3 Maize

Maize presents a more transformation-friendly genome compared to wheat and has seen rapid CRISPR adoption. The first demonstration of CRISPR/Cas9-mediated genome editing in maize targeted the *Argonaute18* gene, laying the groundwork for functional genomics studies [16].

Yield enhancement has focused on editing *ZmLG1* (leaf angle), *ZmWUS* (meristem maintenance), and *ZmCLE* family genes, which control kernel row number and inflorescence architecture [6]. Modifications to *ZmCCT10*, a photoperiod-sensitive gene, have allowed the development of maize lines adaptable to tropical and temperate regions [6].

Nutritional enhancement has been pursued through editing *ZmIPK1*, which reduces phytic acid levels, thus improving micronutrient bioavailability [17]. CRISPR has also enabled resistance to maize lethal necrosis (MLN) and northern corn leaf blight through editing *ZmAGO18* and *ZmNPR1* [18].

4. Progress in pseudocereals

Pseudocereals such as quinoa (*Chenopodium quinoa* Willd.), buckwheat (*Fagopyrum esculentum* Moench), and amaranth (*Amaranthus spp.*) are gaining global attention for their high protein content, balanced amino acid profiles, and gluten-free nature. Despite their nutritional value and resilience to marginal conditions, pseudocereals have historically received limited breeding investment. Recent progress in genome sequencing and gene-editing tools is now unlocking new avenues for their improvement through CRISPR/Cas technologies.

4.1 Quinoa *(Chenopodium quinoa* Willd.*)*

Quinoa has emerged as a model pseudocereal due to its adaptability and rich nutrient profile. The release of the *C. quinoa* genome in 2017 laid the foundation for molecular breeding [19]. However, quinoa is an allotetraploid species with low transformation efficiency, which makes genome editing challenging.

Initial CRISPR efforts in quinoa have targeted genes associated with anti-nutritional factors. For example, editing of *CqAOP1*, which is involved in saponin biosynthesis, has the potential to develop sweet, non-bitter quinoa varieties without postharvest processing the regulatory pathways controlling seed saponin content in quinoa have been mapped to key TFs including TSARL1, which specifically controls seed saponin biosynthesis [20]. Genes involved in stress tolerance, such as heat-shock proteins and transcription factors such as *CqDREB2A*, are also candidates for editing under climate resilience goals.

4.2 Amaranth *(Amaranthus hypochondriacus* L.*)*

Amaranth is a highly adaptable crop with high lysine content, but genomic resources are still limited. The draft genome of *Amaranthus hypochondriacus* L. has enabled preliminary gene annotation [21]. Functional validation of stress-resistance genes using CRISPR remains in the early stages. The use of protoplast systems and transient assays has provided proof-of-concept for CRISPR delivery in amaranth successful CRISPR/Cas9 editing in protoplast systems has been demonstrated, enabling rapid evaluation of gene-editing reagents and trait-targeted events in crops such as potato [22].

Researchers are targeting genes related to seed dormancy, shattering, and plant architecture to facilitate domestication and yield stability. Improvements in transformation protocols and tissue culture systems are critical for the full-scale deployment of CRISPR in this crop.

4.3 Buckwheat *(Fagopyrum esculentum* moench*)*

Buckwheat is widely cultivated in temperate and high-altitude regions and is valued for its antioxidant compounds, such as rutin. Although genetic transformation is difficult in buckwheat, researchers have identified key genes involved in flavonoid biosynthesis (*FePAL*, *FeF3H*) that could be edited to enhance nutraceutical quality. This study reviews breeding approaches in buckwheat related to reducing bitterness and rutin hydrolysis, and provides useful background on flavonoid metabolism and quality traits in pseudocereals [23].

Gene editing in buckwheat is currently limited to experimental systems using *Agrobacterium*-mediated transformation of callus cultures. However, improvements in regeneration and promoter selection may enable stable transformation in the near future. Multiplex genome editing to regulate flowering time and improve lodging resistance is among the goals of ongoing research programs. A summary of key edited traits across these crops is provided below (**Table 1**).

This table summarizes successful applications of CRISPR/Cas technology, highlighting the target gene, the improved agronomic or nutritional trait, and the resulting phenotypic outcome in various crops.

4.4 Challenges and opportunities

The application of CRISPR in pseudocereals is hindered by several factors:

- Poor transformation and regeneration protocols;
- Polyploidy and gene redundancy;
- Limited gene function validation; and

Crop	Target gene(s)	Trait improved	Documented outcome
Rice (Oryza sativa)	OsERF922, Gn1a, GS3, GW2	Blast resistance; grain number, size, and weight.	40% enhanced blast resistance; 12%–18% yield increase.
Wheat (Triticum aestivum)	TaMLO, TaNAC2	Powdery mildew resistance; drought tolerance.	Complete mildew resistance; 15% higher grain weight under drought.
Maize (Zea mays)	ZmLG1, ZmIPK1	Leaf angle (plant architecture); phytic acid reduction.	Improved canopy architecture; enhanced Fe/Zn bioavailability.
Quinoa (Chenopodium quinoa)	CqAOP1	Saponin biosynthesis (bitterness reduction).	Development of "sweet," non-bitter quinoa varieties.
Buckwheat (Fagopyrum esculentum)	FeF3H	Rutin (flavonoid) content for antioxidant activity.	Increased nutraceutical and health-promoting value.

Table 1.
Key CRISPR-edited traits and outcomes in cereals and pseudocereals.

- Lack of publicly available CRISPR vectors optimized for these species.

Despite these challenges, the future is promising. Integration of gene editing with transcriptomics and metabolomics can accelerate trait discovery. Moreover, the DNA-free CRISPR-RNP method holds particular promise for pseudocereals, allowing regulatory-friendly edits without introducing transgenes [24].

5. Integration with omics and AI

The convergence of CRISPR gene editing with advanced omics platforms – genomics, transcriptomics, proteomics, metabolomics – and artificial intelligence (AI) is revolutionizing crop improvement strategies, as depicted in **Figure 2**. These integrative approaches provide a comprehensive understanding of biological systems and significantly enhance the efficiency and precision of genome editing in cereals and pseudocereals.

Omics approaches accelerate CRISPR target discovery. For instance, transcriptomic data help identify drought-responsive transcription factors such as *OsNAC14* and *TaDREB2* [25]. AI-assisted tools such as DeepCRISPR predict gRNA efficiency with >90% accuracy [26].

5.1 Genomics and transcriptomics

High-throughput genome sequencing and transcriptome profiling have greatly accelerated the identification of genes and regulatory networks involved in complex traits such as drought tolerance, nutrient efficiency, and flowering time. For instance, RNA-seq data can identify differentially expressed genes under stress conditions, helping researchers prioritize CRISPR targets . This article provides a comprehensive review of CRISPR/Cas9-based genome editing

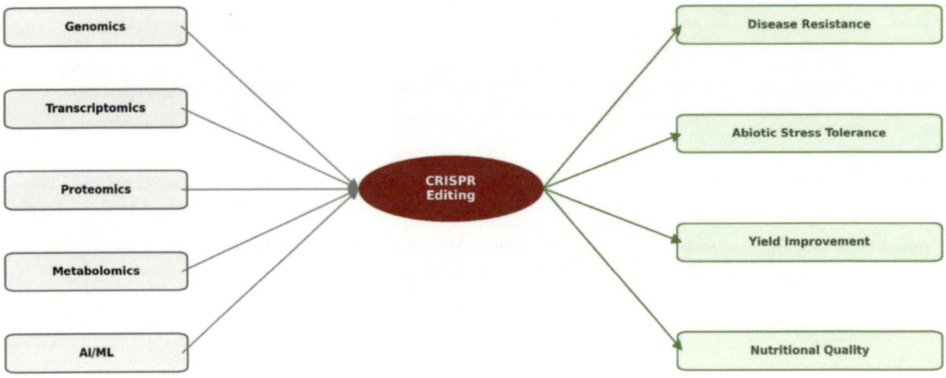

Figure 2.
Integration of multi-omics and artificial intelligence with CRISPR editing for crop improvement. The synergy between genomics, transcriptomics, proteomics, metabolomics, and AI/machine learning (ML) platforms accelerates the discovery of key target genes. This systems biology approach enables precise CRISPR editing to develop improved varieties with enhanced disease resistance, abiotic stress tolerance, yield, and nutritional quality in cereals and pseudocereals.

in plants, covering mechanisms, advances, delivery methods, and applications and is often cited in studies discussing genome editing integration with omics approaches [27].

In rice, the integration of transcriptomics with CRISPR has led to the identification and editing of transcription factors such as *OsNAC14* and *OsbZIP46* for drought tolerance [25]. Similarly, in wheat and maize, genome-wide association studies (GWAS) and expression quantitative trait loci (eQTL) mapping have been used to discover candidate genes linked to yield and resistance traits, which are now amenable to precise editing.

5.2 Proteomics and metabolomics

Proteomic analysis provides insights into posttranslational modifications and protein-protein interactions, revealing the functional impact of gene edits. For example, CRISPR-mediated mutation in *ZmLEC1*, a key seed development gene in maize, was followed by proteome profiling to assess downstream effects on embryo patterning and seed composition [13].

Metabolomics complements gene editing by identifying biochemical pathways affected by edited genes. In quinoa and buckwheat, metabolite profiling helps monitor changes in flavonoid and saponin biosynthesis, enabling targeted nutritional enhancements through CRISPR. Combined transcriptomic and metabolomic analyses provide insights into nutrient synthesis pathways in quinoa seeds [28].

5.3 Artificial intelligence and machine learning

Artificial intelligence and machine learning algorithms are increasingly applied to predict gene function, gRNA efficiency, off-target effects, and phenotypic outcomes of edits. AI-based platforms such as DeepCRISPR and CRISPR-P 2.0 allow researchers to design highly efficient and specific gRNAs, even in complex polyploid genomes [26].

In cereals, ML models trained on genomic and phenotypic datasets have been used to prioritize editing targets for yield improvement and stress tolerance. In pseudocereals, where experimental data are scarce, AI tools are invaluable for predicting gene networks and simulating the impact of putative edits.

5.4 Multi-omics pipelines

The combination of omics layers, termed multi-omics, provides a systems biology framework to study complex traits and gene-environment interactions. For example, integrating genomics, transcriptomics, and metabolomics helped identify regulators of grain protein content in wheat, subsequently validated through CRISPR editing.

Platforms such as Galaxy, OmicsPipe, and iDEP allow researchers to process multi-omics data in an integrative manner, facilitating the selection of optimal CRISPR targets across different species. The integration of these pipelines with AI will usher in a new era of precision breeding tailored to specific environments and markets.

6. Regulatory landscape and public perception

The rapid development of CRISPR and other gene-editing technologies has outpaced the evolution of global regulatory frameworks. The classification, evaluation, and commercialization of genome-edited crops vary significantly across countries, influencing both the deployment of edited cultivars and public acceptance. Understanding the regulatory and societal context is therefore critical for mainstreaming CRISPR applications in cereals and pseudocereals. For instance, the United States exempts DNA-free edits, while the European Union treats them as GMOs [29, 30]. China and India recently introduced science-based, faster approval frameworks.

Public acceptance is improving, especially for edits that enhance nutrition or reduce allergens [31]. Transparent communication and participatory research are essential for building trust in developing regions.

6.1 Global regulatory frameworks

United States: The U.S. Department of Agriculture (USDA) has adopted a relatively permissive stance on gene-edited crops, provided that no foreign DNA is introduced. Several CRISPR-edited crops, including waxy corn and non-browning mushrooms, have been approved without undergoing the full genetically modified organism (GMO) regulatory process [30].

European Union: In contrast, the EU classifies gene-edited organisms as (GMOs under the 2018 ruling of the European Court of Justice. This has created stringent barriers for the approval and release of gene-edited crops [29].

China: China has taken steps toward a more science-based approach. In 2022, the Ministry of Agriculture released new guidelines distinguishing gene-edited crops from transgenic GMOs, indicating a faster approval pathway for CRISPR-derived varieties.

India and South Asia: India, a major agricultural producer, is in the process of finalizing its regulatory approach. In 2022, the Genetic Engineering Appraisal Committee

(GEAC) proposed that gene-edited crops without foreign DNA (especially SDN-1 and SDN-2 edits) would not be subject to GMO regulations. Pakistan and Bangladesh are yet to formalize gene-editing regulations but are closely following India's lead.

These regional differences underscore the need for harmonized, science-driven regulations that distinguish gene editing from transgenesis based on product outcomes rather than process alone.

6.2 Ethical considerations and biosafety

Public concerns about gene editing often echo those raised about GMOs, pertaining to food safety, environmental risks, corporate control, and ethical implications. While CRISPR allows precise edits, unintended off-target mutations remain a theoretical risk, especially in polyploid crops such as wheat and quinoa. Rigorous biosafety assessments, transparency in labeling, and open communication with stakeholders are essential to build trust.

Ethical issues also surround gene editing in minor crops and pseudocereals that are culturally important to Indigenous and smallholder communities. Participatory breeding and equitable access to gene-editing technologies are vital to ensure that innovations do not exacerbate existing inequalities.

6.3 Public perception and communication

Studies show that public acceptance of gene editing is higher than that of GMOs, particularly when edits mimic natural mutations and address consumer-benefiting traits such as nutrition or allergen reduction [31]. However, misinformation and a lack of awareness can still hinder adoption.

Effective science communication strategies are needed to inform consumers, policymakers, and farmers about the safety, benefits, and limitations of CRISPR-edited crops. Collaborative platforms involving scientists, civil society, and the private sector can facilitate dialog and demystify the technology.

6.4 Toward inclusive and adaptive policy

As gene editing moves from the lab to the field, inclusive policy design becomes crucial. Regulatory frameworks must:

- Differentiate between types of edits (e.g., SDN-1 vs. SDN-3);
- Enable fast-track approvals for non-transgenic edits;
- Encourage public–private partnerships in crop development; and
- Support smallholders' access to improved varieties.

Capacity-building in regulatory science, biosafety testing, and intellectual property management is especially important in low- and middle-income countries (LMICs), where many pseudocereals are grown.

7. Future directions and conclusion

The integration of CRISPR and other gene-editing tools into cereal and pseudocereal research has opened unprecedented opportunities for sustainable agriculture, food security, and crop diversification. As we look ahead, several key directions will shape the evolution and impact of these technologies.

7.1 Future directions

7.1.1 Expanding to orphan and underutilized crops

While CRISPR has made major strides in staple cereals such as rice, wheat, and maize, its application in pseudocereals (e.g., quinoa, amaranth, buckwheat) and other underutilized species remains limited. These crops often lack reference genomes and efficient transformation systems, yet they possess immense potential for climate resilience and nutritional enhancement. Future research must invest in building genomic resources and gene-editing protocols for these species to unlock their full potential.

7.1.2 Precision breeding for complex traits

Traits governed by multiple genes, such as drought tolerance, nutrient use efficiency, and grain quality, require multiplex editing strategies. Advances in base editing, prime editing, and RNA-guided epigenetic modifications are enabling scientists to fine-tune gene expression without introducing double-strand breaks or foreign DNA. The convergence of these approaches with AI-driven gene discovery will pave the way for ultra-precise crop improvement.

7.1.3 CRISPR for adaptation to climate change

With the increasing frequency of extreme weather events and shifting disease pressures, there is an urgent need to develop climate-resilient crop varieties. CRISPR can expedite the breeding of genotypes with enhanced heat, salinity, and drought tolerance. The strategic targeting of specific genes for major abiotic stresses is summarized in **Figure 3**. Recent efforts in editing genes such as *OsDREB1*, *TaHKT1*, and *ZmNAC111* are indicative of this trend. Future work must align gene-editing goals with climate-smart agriculture frameworks.

7.1.4 Democratizing gene editing

To ensure equitable benefits from gene-editing innovations, there is a need to lower technical and policy barriers in low- and middle-income countries. This includes the following:

- Developing low-cost and tissue culture-independent editing systems;
- Sharing germplasm, genome assemblies, and transformation protocols through open-source platforms; and

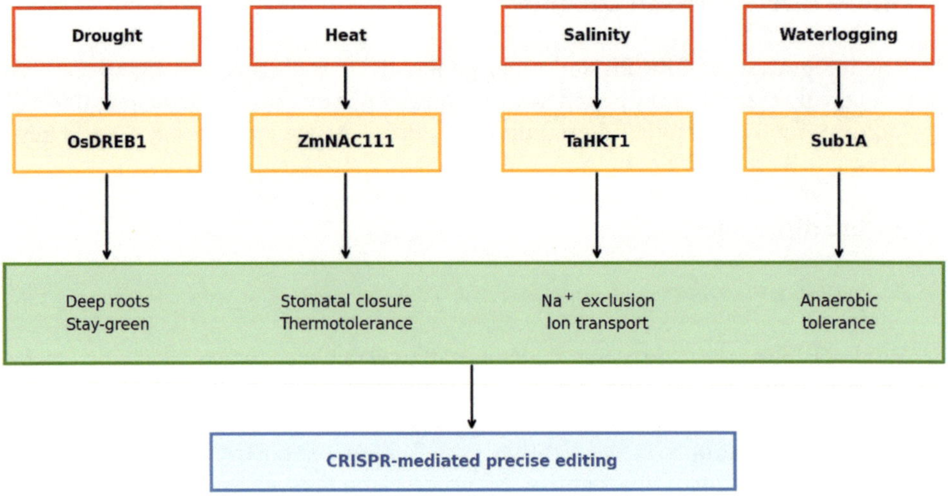

Figure 3.
Future prospects of CRISPR for developing climate-resilient crops. Key target genes for editing in cereals to confer tolerance to major abiotic stresses are shown. Examples include OsDREB1 for drought, ZmNAC111 for heat, TaHKT1 for salinity, and Sub1A for waterlogging tolerance. The corresponding physiological traits, such as deep root systems, Na$^+$ exclusion, and aerenchyma formation, contribute to enhanced resilience and stable yields under adverse climatic conditions.

- Facilitating regional regulatory harmonization to ease the release and trade of edited crops.

7.1.5 Social acceptance and responsible innovation

A multidisciplinary approach involving scientists, ethicists, policymakers, farmers, and consumers is crucial to building socially acceptable gene-editing pathways. Public trust can be fostered through transparency in research, proactive stakeholder engagement, and inclusive governance models.

8. Conclusion

CRISPR and other gene-editing technologies have transformed the landscape of plant breeding by offering precision, speed, and versatility unmatched by conventional methods. In cereals and pseudocereals, these tools are being used not only to enhance yield and stress tolerance but also to improve nutritional value and functional traits.

Despite regulatory challenges and technical limitations, the momentum behind gene editing is undeniable. As more genomes are sequenced, editing platforms are refined, and data is integrated through AI and omics, we are poised to enter an era of truly rational crop design.

By embracing a forward-looking, inclusive, and ethically grounded approach, CRISPR can become a cornerstone of sustainable agriculture, meeting global food demands while preserving biodiversity and empowering local communities.

Artificial Intelligence assistance disclosure

This chapter benefited from the use of an AI language model (ChatGPT, OpenAI) to assist with improving grammar, structure, and clarity during the writing process. All intellectual content, data interpretation, and scientific conclusions were fully determined and verified by the authors. The authors take full responsibility for the accuracy and integrity of the content presented in this chapter.

Author details

Akash Fatima[1*], Zareen Sarfraz[2], Muhammad Ali Raza[3,4], Sajad Hussain[3,5], Akhtar Hameed[6], Fawad Zafar Ahmad Khan[6], Sehrish Ijaz[1] and Ummara Waheed Khan[1]

1 Institute of Plant Breeding and Biotechnology, MNS University of Agriculture, Multan, Pakistan

2 Agricultural Genomics Institute at Shenzhen (AGIS), Chinese Academy of Agricultural Sciences (CAAS), Shenzhen, China

3 National Research Center of Intercropping, The Islamia University of Bahawalpur, Bahawalpur, Pakistan

4 Gansu Academy of Agricultural Sciences, Lanzhou, China

5 College of Life Sciences, Shandong Agricultural University, Tai'an, China

6 Institute of Plant Protection, MNS University of Agriculture, Multan, Pakistan

*Address all correspondence to: akash.fatima@mnsuam.edu.pk

References

[1] Borrelli VMG, Brambilla V, Rogowsky P, Marocco A, Lanubile A. The enhancement of plant disease resistance using CRISPR/Cas9 technology. Frontiers in Plant Science. 2018;**9**:1245. DOI: 10.3389/fpls.2018.01245

[2] Bortesi L, Fischer R. The CRISPR/Cas9 system for plant genome editing and beyond. Biotechnology Advances. 2015;**33**(1):41–52. DOI: 10.1016/j.biotechadv.2014.12.006

[3] Chen K, Wang Y, Zhang R, Zhang H, Gao C. CRISPR/Cas genome editing and precision plant breeding in agriculture. Annual Review of Plant Biology. 2019;**70**:667–697. DOI: 10.1146/annurev-arplant-050718-100049

[4] Li J, Meng X, Zong Y, et al. Gene replacements and insertions in rice by intron targeting using CRISPR–Cas9. Nature Plants. 2016;**2**:16139

[5] Jinek M, Chylinski K, Fonfara I, Hauer M, Doudna JA, Charpentier E. A programmable dual-RNA–guided DNA endonuclease in adaptive bacterial immunity. Science. 2012;**337**(6096):816–821. DOI: 10.1126/science.1225829

[6] Tang X, Liu G, Zhou J, et al. A large-scale whole-genome sequencing analysis reveals highly specific genome editing by both Cas9 and Cpf1 (Cas12a) nucleases in rice. Genome Biology. 2020;**21**:235

[7] Anzalone AV, Randolph PB, Davis JR, et al. Search-and-replace genome editing without double-strand breaks or donor DNA. Nature. 2019;**576**(7785):149–157. DOI: 10.1038/s41586-019-1711-4

[8] Woo JW, Kim J, Kwon SI, et al. DNA-free genome editing in plants with preassembled CRISPR-Cas9 ribonucleoproteins. Nature Biotechnology. 2015;**33**(11):1162–1164. DOI: 10.1038/nbt.3389

[9] Wang M, Lu Y, Botella JR, Mao Y, Hua K, Zhu JK. Gene targeting by homology-directed repair in rice using a geminivirus-based CRISPR/Cas9 system. Molecular Plant. 2017;**10**(7):1007–1010. DOI: 10.1016/j.molp.2017.04.007

[10] Jin ML, Ding DY. Multiplex CRISPR-Cas9 editing of yield-related genes in rice. Rice Genomics and Genetics. 2025;**16**(3):140–149. DOI: 10.5376/rgg.2025.16.0013

[11] Hu X, Cui Y, Dong G, Feng A, Wang D, Zhao C, Zhang Y, Hu J, Zeng D, Guo L, Qian Q. Using CRISPR-Cas9 to generate semi-dwarf rice lines in elite landraces. Scientific Reports. 2019;**9**(1):19096. DOI: 10.1038/s41598-019-55757-9

[12] Liang Z, Chen K, Li T, Zhang Y, Wang Y, Zhao Q, Liu J, Zhang H, Liu C, Ran Y, Gao C. Efficient DNA-free genome editing of bread wheat using CRISPR/Cas9 ribonucleoprotein complexes. Nat Commun. 2017;**8**:14261. DOI: 10.1038/ncomms14261

[13] Zhang Q, Chen Q, Wang S, Hong Y, Wang Z. Exploration of drought-tolerant gene function by CRISPR/Cas9 technology in rice. Plant Cell Reports. 2020;**39**:799–808. DOI: 10.1007/s00299-020-02531-1

[14] Sánchez-León S, Gil-Humanes J, Ozuna CV, Giménez MJ, Sousa C,

Voytas DF, Barro F. Low-gluten, nontransgenic wheat engineered with CRISPR/Cas9. Plant Biotechnology Journal. 2018;**16**(4):902–910. DOI: 10.1111/pbi.12837

[15] Abdallah NA, Elsharawy H, Abulela HA, Thilmony R, Abdelhadi AA, Elarabi NI. Multiplex CRISPR/Cas9-mediated genome editing to address drought tolerance in wheat. GM Crops & Food. 2022;**13**(1):1–17. DOI: 10.1080/21645698.2022.2120313

[16] Svitashev S, Young JK, Schwartz C, Gao H, Falco SC, Cigan AM. Targeted mutagenesis, precise gene editing, and site-specific gene insertion in maize using Cas9 and guide RNA. Plant Physiology. 2015;**169**(2):931–945. DOI: 10.1104/pp.15.00793

[17] Shukla VK, Doyon Y, Miller JC, et al. Precise genome modification in the crop species Zea mays using zinc-finger nucleases. Nature. 2009;**459**(7245): 437–441. DOI: 10.1038/nature07992

[18] Qi X, Zhang C, Zhu J, Liu C, Huang C, Li X, Xie C. Genome editing enables next-generation hybrid seed production technology. Molecular Plant. 2020;**13**(10):1472–1481. DOI: 10.1016/j.molp.2020.06.003

[19] Jarvis DE, Ho YS, Lightfoot DJ, Schmöckel SM, Li B, Borm TJA, Ohyanagi H, Mineta K, Michell CT, Saber N, Kharbatia NM, Rupper RR, Sharp AR, Dally N, Boughton BA, Woo YH, Gao G, Schijlen EGWM, Guo X, Momin AAMI, Negrão S, Al-Babili S, Gehring CA, Roessner U, Jung C, Murphy K, Arold STA, Gojobori T, van der Linden CG, van Loo EN, Jellen EN, Maughan PJ, Tester M. The genome of Chenopodium quinoa. Nature. 2017;**542**(7641):307–312. DOI: 10.1038/nature21370

[20] Trinh MDL, Visintainer D, Günther J, Østerberg JT, Rodrigues da Fonseca RA, Fondevilla S, Moog MW, Luo G, Nørrevang AF, Crocoll C, Nielsen PV, Jacobsen S-E, Wendt T, Bak S, López-Marqués RL, Palmgren M. Site-directed genotype screening for elimination of antinutritional saponins in quinoa seeds identifies TSARL1 as a master controller of saponin biosynthesis selectively in seeds. Plant Biotechnology Journal. 2024;**22**(8):2216–2234. DOI: 10.1111/pbi.14340

[21] Lightfoot DJ, Jarvis DE, Ramaraj T, Lee R, Jellen EN, Maughan PJ. Single-molecule sequencing and Hi-C-based proximity-guided assembly of amaranth (Amaranthus hypochondriacus) chromosomes provide insights into genome evolution. BMC Biology. 2017;**15**(1):74. DOI: 10.1186/s12915-017-0412-4

[22] Rather GA, Ayzenshtat D, Teper-Bamnolker P, Kumar M, Forotan Z, Eshel D, Bocobza S. Advances in protoplast transfection promote efficient CRISPR/Cas9-mediated genome editing in tetraploid potato. Planta. 2022;**256**(1):14. DOI: 10.1007/s00425-022-03933-z

[23] Suzuki T, Morishita T, Noda T, Ishiguro K, Otsuka S, Katsu K. Breeding of buckwheat to reduce bitterness and rutin hydrolysis. Plants. 2021;**10**(4):791. DOI: 10.3390/plants10040791

[24] Tripathi L, Dhugga KS, Ntui VO, Runo S, Syombua E, Muiruri S, Wen Z, Tripathi JN. Genome Editing for Sustainable Agriculture in Africa. Frontiers in Genome Editing. 2022;**4**:876697. DOI: 10.3389/fgeed.2022.876697

[25] Li C, Zong Y, Wang Y, et al. Expanded base editing in rice and wheat

using a Cas9 adenosine deaminase fusion. Genome Biology. 2018;**19**:59

[26] Chuai G, Ma H, Yan J, Chen M, Hong N, Xue D, Zhou C, Zhu C, Chen K, Zhang J, et al. DeepCRISPR: optimized CRISPR guide RNA design by deep learning. Genome Biology. 2018;**19**(1):80. DOI: 10.1186/s13059-018-1459-4

[27] Wada N, Ueta R, Osakabe Y, Osakabe K. Precision genome editing in plants: state-of-the-art in CRISPR/Cas9-based genome engineering. BMC Plant Biology. 2020;**20**(1):234. DOI: 10.1186/s12870-020-02385-5

[28] Yang J, Wang Y, Sun J, Li Y, Zhu R, Yin Y, Wang C, Qin L. Metabolome and transcriptome association analysis reveals mechanisms of nutrient composition synthesis in quinoa (Chenopodium quinoa Willd.) seeds. Foods. 2024;**13**(9):1325. DOI: 10.3390/foods13091325

[29] Callaway E. CRISPR plants now subject to tough GM laws in European Union. Nature. 2018;**560**:16. DOI: 10.1038/d41586-018-05814-6

[30] Waltz E. Gene-edited CRISPR mushroom escapes US regulation. Nature. 2016;**532**:293. DOI: 10.1038/nature.2016.19754

[31] Shew AM, Nalley LL, Snell HA, RM Jr N, Dixon BL. CRISPR versus GMOs: Public acceptance and valuation. Global Food Security. 2018;**19**:71–80. DOI: 10.1016/j.gfs.2018.10.005

Chapter 2

Screening of Genomic Sequences of Allergens and Their Homologs in the Genomes of Pseudocereals *Fagopyrum* spp. Moench and *Amaranthus* spp. L.

Silvia Farkasová, Jana Žiarovská and Eva Ivanišová

Abstract

Pseudocereals are increasingly consumed as healthier alternatives to traditional cereals due to their high micronutrient and phytonutrient content and gluten-free nature. However, information regarding their allergenic safety remains limited. Hypersensitivity reactions caused by homologous proteins from different dietary sources represent an important clinical issue. Based on the hypothesis that sequence homology may exist among cereal, pseudocereal, and pollen allergen - particularly PR-10 proteins and profilins, this study combined in silico and polymerase chain reaction (PCR) approaches. The study summarizes available genomic data on allergens from selected pseudocereals, specifically *Fagopyrum* spp. and *Amaranthus* spp. In silico analysis was performed using BLAST to compare sequences of major cereal and pollen allergens, including Bet v 1 and profilins, with pseudocereal genomic data. In addition, key pseudocereal allergens were analyzed for homology with the genomes of commonly consumed cereals. Selected results were validated by PCR analysis. The findings demonstrate the presence of Bet v 1 and profilin homologs in the genomes of the studied pseudocereals. Furthermore, homologs of the major buckwheat allergen Fag e 1 were identified in wheat, barley, oats, and amaranth. These results provide evidence for potential cross-reactivity among cereal, pseudocereal, and pollen allergens. Improved understanding of these relationships may support the development of more accurate diagnostic tools and personalized management strategies for food-allergic individuals.

Keywords: allergens, gene homologs, *in silico*, pseudocereals, PCR, *Fagopyrum* spp, *Amaranthus* spp

1. Introduction

Cereals are staple foods that satisfy over 50% of the world's energy demand, with wheat, maize, and rice being the most significant. Recently, there has been a growing interest in pseudocereals among consumers, as these crops offer suitable alternatives to traditional cereals from a nutritional standpoint, despite not being classified as cereals taxonomically. Pseudocereals include, for instance, buckwheat and amaranth. These crops are often promoted as healthier alternatives to cereals, being rich in micronutrients and phytonutrients, and are also considered gluten-free, making them suitable for individuals with gluten-related disorders. Such crops frequently feature in the diets of those suffering from celiac disease [1].

Fagopyrum spp.

Buckwheat flourishes in nutrient-poor soils and in regions where most cereals cannot thrive. Additionally, the crop exhibits resistance to pests and diseases. These traits, along with its short growing season, provide buckwheat with exceptional potential for both organic and conventional agriculture [2, 3].

Buckwheat is a plant that belongs to the genus *Fagopyrum* Mill. and the family *Polygonaceae*, Juss., which comprises 17 species [4]. Of these, *Fagopyrum esculentum* Moench. and *Fagopyrum tataricum* L. are diploid species that are commonly cultivated for human consumption ($2n = 16$) [5, 6]. This herb is native to Central Asia and is known for its high nutritional value, featuring greater protein and fatty acid content compared to cereals [7].

One of the key benefits of buckwheat is its gluten-free nature, and numerous scientific studies have explored the various medicinally active compounds found in it. The medicinal properties of buckwheat are attributed to the presence of tocopherols (0.3%–0.4%), phytosterols (4%–6%), and squalane in the lipid component of the grains, along with a rich array of bioactive flavonoids [8]. Buckwheat (*Fagopyrum* spp.) has attracted significant interest from the global scientific community due to its nutritional and pharmaceutical properties. Its gluten-free nature, balanced amino acid composition of the grain protein, and high levels of antioxidants such as rutin render buckwheat an important crop with considerable nutritional benefits [9, 10].

Buckwheat may be beneficial in addressing several chronic human diseases, including hypertension, obesity, cardiovascular disease, and gallstone formation [11]. Buckwheat grains contain 100–125 mg/g protein, 650–750 mg/g starch, 20–25 mg/g fat, and 20–25 mg/g minerals. The low content of prolamins and the absence of α-gliadin, which is crucial for the gluten-free nature of buckwheat grain proteins, establish it as a healthy alternative to gluten-containing cereals for those with gluten-related disorders [12].

Amaranthus spp.

Recently, amaranth has gained more recognition for its nutritional benefits, especially regarding the quality of its grain proteins. Amaranth has a higher fiber content (8%) than most cereal grains. Its crude protein content ranges from 12.5% to 22.5% on a dry matter basis, which is 50% higher than that found in cereals like wheat (*Triticum aestivum* L.), rice (*Oryza sativa* L.), or maize (*Zea mays* L.). Lysine, often the limiting amino acid in other cereal crops, is present in relatively high levels in amaranth grains. The mineral content of these grains, including iron (Fe), magnesium (Mg), and calcium (Ca), is particularly high, making amaranth flour an excellent candidate to substitute for wheat flour [13, 14].

Amaranth belongs to the *Amaranthaceae* family, which comprises over 60 genera and approximately 800 species of dicotyledonous herbaceous plants, both annual and perennial. The three cultivated species of the genus Amaranthus produce relatively large inflorescences, often yielding more than 50,000 edible seeds per plant. These species are *A. hypochondriacus*, native to Mexico; *A. cruentus*, originating from Guatemala; and *A. caudatus*, which is indigenous to Peru and other Andean countries. Amaranths thrive exceptionally well in hot, humid regions of Africa, Southeast Asia, southern China, and India [15]. Traditionally, various species of amaranth are cultivated either for their grain or as leafy vegetables [16].

1.1 Food allergies

The earliest documented instances of adverse health issues, such as allergies and asthma, can be traced back over 3,000 years to ancient Egypt, Rome, and China [17]. The mechanisms underlying allergic reactions involve complex interactions between proteins and the immune system, rendering them challenging to predict [18].

Allergy is currently the leading chronic disease in Europe, affecting more than 20% of the population. In the near future, its prevalence is even predicted to increase by 40%. Around 17 million of the European population suffer from food allergies, and 3.5 million among these patients are under the age of 25. The geographic variability of major allergens is reported to reflect a potential interaction of genetic factors, age, cultural and dietary habits, together with the time of first exposure to allergenic products [19], and other factors such as socioeconomic class [20], urbanization – the so-called hygiene hypothesis [21], ethnicity [22], and migration patterns [23]. Allergens contain B-cell epitopes that can bind to immunoglobulin E (IgE), as well as T-cell epitopes capable of triggering a type 2 T-lymphocyte response [24]. IgE-mediated food allergy affects approximately 3% of the population and significantly impacts the daily activities of those affected. Symptoms can manifest not only in the gastrointestinal tract but also across other organ systems. When dietary antigens interact with IgE, cross-linking occurs, leading to the binding of IgE to mast cells and basophils. This activation triggers the release of granules containing preformed inflammatory mediators, such as histamine, as well as the synthesis and release of additional inflammatory mediators, including leukotrienes, proteases like tryptase, inflammatory cytokines such as IL-4, and chemotactic molecules. Mast cells and basophils become activated within minutes of IgE binding, resulting in a rapid allergic reaction, with symptoms appearing shortly after contact with the allergen [25, 26].

Cross-allergic responses present challenges for patients, as they are induced by homologous proteins. Allergen isoforms were initially defined as distinct molecular variants of the same allergen that exhibit significant cross-reactivity (IgE). The updated terminology specifies that isoforms are allergens originating from the same species, exhibiting analogous molecular structures and biological activities, and sharing a minimum of 67% sequence similarity [17, 27, 28]. The rising incidence of allergic sensitization has led to the prevalence of responses to various foods containing homologous proteins. The preservation of these proteins in diverse biological tissues influences cross-reactivity. Some foods, like peanuts, can sensitize individuals and trigger responses upon oral exposure (type 1 allergy), and may also cause reactions to related foods, such as legumes [29].

2. Allergens of *Fagopyrum* spp.

Buckwheat allergy is an IgE-mediated hypersensitivity reaction that can lead to severe symptoms, including anaphylactic shock. The allergens associated with buckwheat are primarily the storage proteins found in its seeds (see **Tables 1** and **2**) [30].

Electrophoretic immunoblotting revealed distinct binding patterns between IgE antibodies and these storage proteins among the 37 patients under study. The proteins identified include those of sizes 17, 50, and 100 kDa; 8–9 kDa; 24 kDa; 22 kDa; 36 kDa; 39–40 kDa; and 70–72 kDa, as well as 14, 18, and 16 kDa. The 22 kDa protein, which exhibited binding activity with nearly all blood sera from the patients, is regarded as the principal allergic protein of buckwheat (Fag e 1) [30].

Several clinically significant allergenic proteins have been identified in buckwheat, including Fag e 1, Fag e 2, and Fag e 3, as well as in buckwheat tartar (Fag t 1, Fag t 2, and Fag t 3). Researchers have reported clinically relevant cross-reactivity between buckwheat and peanuts, latex, coconut, quinoa, and poppy seeds. The estimated incidence of severe allergic reactions to buckwheat, including anaphylaxis, is approximately 0.1–0.01 cases per 100,000 individuals [31]. Studies by Sammut et al. [32] and Badiu et al. [33] indicate a rising number of patients with buckwheat allergy in Europe. Given the growing popularity of foods potentially containing buckwheat, such as ethnic and health foods, it is probable that buckwheat allergy will become increasingly prevalent.

The three-dimensional model developed by Sordet et al. [34] for the 13 S globulin allergen (Fag e 1) of buckwheat (*Fagopyrum esculentum* Moench.) consists of three protomers with a cluster motif, arranged in a homotrimer around a three-fold axis of symmetry. They identified 11 continuous IgE-binding epitope peptides on the

Name of the allergen	Characteristics	mRNA/DNA sequence in NCBI database
Fag e 1	13 S globulin	AB828119.1
Fag e 2	Albumin	DQ304682.1
Fag e 3	Vicilin	EF532322.1
Fag e 4	Antimicrobial peptide	x
Fag e 5	Globulin	AY536051.1
Fag e 10 kDa	Amylase inhibitor	AB055892.1
Fag e TI	Trypsin inhibitor	x

Table 1.
Allergens of Fagopyrum esculentum *Moench (Allergome database, 2025).*

Name of the allergen	Characteristics	mRNA/DNA sequence in NCBI database
Fag t 1	Globulin	DQ849083.1
Fag t 2	Albumin	HQ829975.1
Fag t 24 kDa	Unknown	x
Fag t 6	Oleosin	MW567728.1

Table 2.
Allergens of Fagopyrum tataricum *L. (Allergome database, 2025).*

molecular surface of the buckwheat 13 S globulin allergen. Notably, half of these identified epitope peptides remain unchanged in their core structure even after hydrolysis by digestive proteases, which supports the allergenicity of the 13 S globulin. Moreover, several of these epitopes have similar sequences to IgE-binding epitopes previously characterized in soybean 11 S globulins, potentially explaining the IgE-binding cross-reactions observed in Western blot experiments between soybean and buckwheat. Khan et al. [35] reported that the structure of the 13 S allergen comprises polypeptide subunits and that tandem repeats within these polypeptides are significant for resistance to the digestive enzyme trypsin.

A study [36] examined the 8S storage globulin from buckwheat seeds, which shares structural similarities with the vicilin family of storage proteins. The findings indicated that the expression of the 8 S globulin gene precedes that of the 13 S globulin (the main storage protein of buckwheat) and commences early in buckwheat seed development (9–11 days after flowering). The expression of the 8 S globulin gene continues to accumulate during seed development, contributing approximately 7% of the total seed protein. A partial mRNA was also isolated that, when transcribed into a cDNA sequence, exhibited high homology to DNA sequences encoding vicilin-like storage proteins in various plant species.

3. Allergens of *Amaranthus* spp.

The Allergome database does not list any allergens for commonly grown amaranth species, such as *A. hypochondriacus*, *A. caudatus*, and *A. cruentus*. However, two allergens (see **Table 3**) have been identified for *A. retroflexus* according to the Allergome database (2025).

Globulin 11 S, or amaranthin, is the principal fraction in amaranth protein isolates; it has a molecular weight of 56 kDa and consists of three subunits. The 7 S globulin is present in smaller quantities than the 11 S globulin and is less well-studied, comprising four subunits with molecular weights of 66, 52, 38, and 16 kDa. Albumins in amaranth are rich in methionine and have a molecular weight of 18 kDa. Prolamins are less abundant in amaranth, consisting of three fractions with molecular weights ranging from 52 to 54 kDa, 33 to 34 kDa, and 22 to 27 kDa, linked by sulfide bonds [37].

Reports have indicated allergies to *Amaranthus spinosus* pollen, despite its classification as a wild weed. A survey revealed that amaranth pollen ranks among the most commonly found pollen types in Thailand. Furthermore, patients with allergic rhinitis in Thailand have exhibited sensitivity to amaranth pollen extract. Nonetheless, only one allergenic component from amaranth pollen, named Ama r 2 from *A. retroflexus*, has been identified and sequenced. Ama r 2 is a 14.6 kDa profilin-like protein composed of 133 amino acids [38]. To date, there are no documented cases of food allergies related to cultivated amaranth seeds.

Name of the allergen	Characteristics	mRNA/DNA sequence in NCBI datase
Ama r 1	"Ole e 1-like" protein	KR870437.1
Ama r 2	Profilin	FJ899746.1

Table 3.
Allergens of Amaranthus retroflexus *L. (Allergome, 2025).*

A rare case of an allergic reaction induced by *Amaranthus paniculatus* flour was investigated by Kasera et al. [39] in a 60-year-old female patient in India. The patient's serum exhibited a high concentration of IgE (416 IU/ml) against Amaranthus grains. The IgE-binding components of extracts prepared from both raw and roasted Amaranthus grains were analyzed using Western blotting with the patient's sera, revealing protein fractions of sizes 13, 16, 18, 20, 22, 24, 27, 36, 38, 41, 43, 46, 54, 57, 67, and 72 kDa. The ELISA assay indicated protein fractions of sizes 14, 28, and 66 kDa. A study by Cárdenas-Torres et al. [40] investigated the allergenicity of albumins and globulins from *A. hypochondriacus*. The experiment utilized a mouse model, and specific IgE and IgG antibodies were assessed for all tested proteins. The results demonstrated that mice were sensitized to the non-extruded albumin fraction of amaranth grains. A literature search revealed no other documented cases of allergy to amaranth, which is generally considered a hypoallergenic food. This suggests that while *A. hypochondriacus* may pose an allergenic risk, amaranth consumption remains largely safe for most individuals. Further research is necessary to explore any potential cross-reactivity and to enhance understanding of the immunological responses elicited by these proteins.

4. Study of genomic sequences of allergens using molecular genetics and bioinformatics tools

4.1 Bioinformatics and molecular approaches in allergy research

Bioinformatics tools play a crucial role in identifying and analyzing plant allergens by assessing sequence and structural similarities between novel proteins and known allergens (see **Figure 1**). These tools, often used in conjunction with databases like SDAP (Structural Database of Allergenic Proteins) and Allergome, help predict potential allergenicity and cross-reactivity. Regardless of the gene source of a protein, shared amino acid sequence similarities to other known proteins can provide valuable information, such as the level of structural relatedness to proteins with a known history of safe use or to proteins that are known to be allergens. Typically, sequence comparison searches are conducted using various

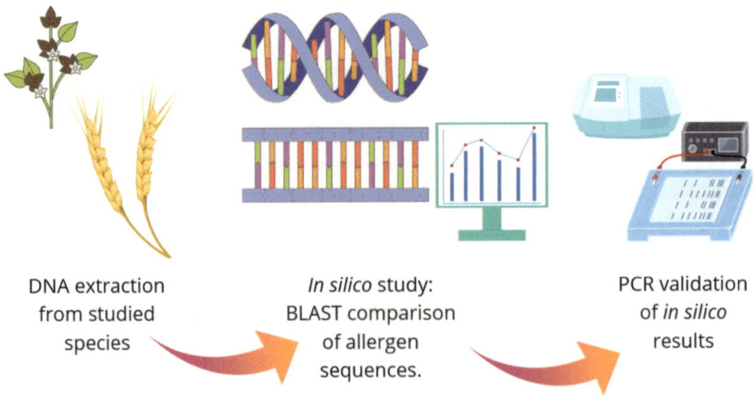

Figure 1.
In silico *and PCR methodology in allergen gene sequence studies.*

algorithms such as Basic Local Alignment Search Tool (BLAST) or FAST-All (FASTA) [41–43], which were originally designed to find evolutionary relationships between protein sequences based on regions of shared amino acid identity and similarity.

In allergy-related research, scientists employ a combination of general and targeted bioinformatics methods. Traditional approaches include sequence comparison, alignment, motif and pattern identification, and the analysis of various protein sequence features [44]. Homologous proteins often possess similar three-dimensional structures, and such structural resemblance can suggest functional similarity. Proteins that serve analogous biological roles frequently display similarities in their primary amino acid sequences. Therefore, the initial step in analyzing a novel protein typically involves identifying homologous sequences within public databases.

The FASTA and BLAST programs are widely used for this purpose. A major category of bioinformatics tools focuses on pairwise sequence alignment, evaluating similarities between a query sequence and a set of target sequences. Establishing accurate local or global multiple sequence alignments is computationally intensive but essential in molecular biology. One of the most commonly used tools for multiple sequence alignment is ClustalW [45, 46].

Efficient storage and management of data have become very important because of the incessant accumulation of molecular and clinical data on allergens. Therefore, allergen databases represent crucial resources for basic allergy research, as they are involved in the archival of available allergen knowledge. Many allergen-specific databases have been developed in the past few years, although they differ from each other with respect to their objectives, the type of data archived, the accessibility of contents, the level of annotation, and their applications [46, 47].

4.2 Polymerase chain reaction in allergen detection

The detection of allergens or traces of allergens in food is a crucial aspect from the perspective of food safety. Various methodologies for detecting allergens have been developed, including the ELISA test, which is based on proteins. Recently, DNA-based tests have also become standard practice in allergen detection [48], as they are mostly unaffected by heat treatment in food, unlike protein-based tests [49]. Real-time polymerase chain reaction (qPCR) has also been recognized as a powerful indirect method for assessing allergens by measuring the expression of allergen-encoding genes. This technique provides precise, sensitive, and quantitative detection of food allergens [50, 51]. Its effectiveness has been demonstrated in numerous studies. For instance, PCR assays have been developed for detecting allergens in shellfish, crustaceans, and lobsters [52]; tree nuts [53]; and peanuts [54, 55]. It has also been applied to chestnuts and pistachios [56]; walnuts and pecans [57]; soy and hazelnuts [58], among others.

4.3 Molecular markers in plant allergen research

The development and application of molecular markers began in the 1980s and gained momentum with the advent of PCR-based technologies. These markers have since become essential tools in plant genomics and molecular breeding. Molecular markers, which are typically nucleic acid sequences showing polymorphism, reveal genetic diversity caused by point mutations, insertions, or deletions at primer binding sites. PCR-based protocols using such markers are now standard in plant genetic studies [59]. These methodologies can be based on coding gene families [60, 61] or non-coding parts of the DNA [62]. They can target specific parts of the DNA or arbitrarily amplified motifs [63]. Importantly, these methods can target genes associated with allergenic proteins. Marker methodologies utilizing genes associated with allergenic proteins allow the design of primers that can detect conserved regions of genes present in various plant species. For instance, the BBAP method effectively identifies conserved genes related to the Bet v 1 allergen. Bet v 1 is the main birch tree allergen. Bet v 1 is a protein belonging to the PR-10 (pathogen-related) protein family. These proteins can be found in many plant species and are involved in cross-reactions in sensitive patients [64].

Profilin genes can also be used as a basis for PCR marker development. Profilins generally display a high degree of sequence conservation suggesting evolutionary stability across species, though there is also evidence of genetic variability. Variations in profilin gene sequences may influence both the functional properties and regulatory mechanisms of these proteins. This genetic diversity could lead to the formation of distinct profilin isoforms, or isovariants, which might differ in their allergenic potential across plant species. Despite these variations, profilin isoforms among plants tend to exhibit relatively low sequence divergence [65]. This conserved nature allows them to act as cross-reactive allergens, meaning sensitization to profilin from one plant source can lead to allergic reactions to others. As a result, profilins are considered major contributors to the phenomenon of cross-reactivity observed in allergies involving pollen from monocots and dicots, as well as in a wide range of plant-based foods [66].

5. Screening of homologs related to cereal and pollen allergens in the genomes of *Fagopyrum* spp. and *Amaranthus* spp.

This study aims to investigate the presence of allergen DNA sequence homologs across various plant species to assess their potential role in cross-reactivity and allergenic sensitization. Specifically, the research focuses on identifying conserved sequences among known allergens and their homologs, using bioinformatics and molecular tools, to enhance the prediction and risk assessment of allergenicity in food sources. The significant allergens found in wheat were compared with the genomes of buckwheat and amaranth species. Bioinformatics and PCR methodologies were employed as an initial screening to identify potential cross-reactions, based on the premise that clinically significant cross-reactivity may occur among cereal grains and grasses [67]. Additionally, the clinically relevant buckwheat allergen Fag e 1 was compared with the genomic records of commonly consumed cereals. Marker techniques involving the Bet v 1 and profilin genes were also utilized, as they can provide insights into the presence of homologous genes within the genomes of the studied pseudocereals.

5.1 Comparative analysis of key wheat allergen sequences in relation to the genomes of *Amaranthus* spp. and *Fagopyrum* spp., along with PCR validation

For the analysis, five major wheat allergens were selected with known IgE epitopes: Tri a 14 – lipid transfer protein (*cDNA: X63669.1, forward primer 5′–3′:

ATCCGTCAAGCTGGTGC; reverse primer 5′–3′: ATGGTACAAACTCTCAACTCT), Tri a 15 – α-amylase inhibitor (*cDNA: FN435985.1; forward primer 5′–3′: AGTGGTCCTTGGAGTTGGTGC, reverse primer 5′–3′: CGCCGAGCTCCTGATAAACA), Tri a 19 – ω-5-gliadin (*DNA: AB181300.1; forward primer 5′–3′: CAACAATWHYCCACAACAACAA, reverse primer 5′–3′: ACAACGATGATTCACCCGTCT), Tri a 33 – serpin (*mRNA: AJ245879.1; forward primer 5′–3′: ACTAAGGCCGCGGAAGTTAC, reverse primer 5′–3′: CTTCTGCCTTGGGATATGGC), and Tri a 37 - purothionin (*DNA: X96446.1; forward primer 5′–3′: GCAGGCGTCTGTAGGTGTAA, reverse primer 5′–3′: ACCTGCGTTGACCATGTAGT; allergome.org; uniprot.org). Based on the accession number * in the NCBI (National Centre for Biotechnology Information) database, the nucleotide sequence of the allergens and their corresponding protein sequences were retrieved. This sequence was then compared with the genomes of *Amaranthus* spp. and *Fagopyrum* spp. using the BLAST (discontiguous megablast) algorithm within the NCBI database. From the information obtained through the BLAST analysis, primers were designed to detect conserved gene regions of the selected allergens by PCR methodology.

The temperature profile for the PCR analysis of homolog detection of the allergens was as follows: initial denaturation at 95°C, followed by 35 cycles comprising denaturation at 95°C for 15 seconds, primer annealing at 65°C for 15 seconds, and polymerization at 72°C for 20 seconds (except for Tri a 37, which required 25 seconds); final polymerization was conducted at 72°C for 10 minutes. Individual reactions were performed in a volume of 10 μL, utilizing MasterMix Robust HS Elizyme 2x (Elizabeth Pharmacon) with 400 nM of each primer. The DNA concentration for each reaction was set at 20 ng/μL (except for Tri a 37, which was at 80 ng/μL). The biological material consisted of multiple cultivars of *Fagopyrum esculentum* (Špačínska 1, Jana C1, Pyra, Aďa) and *Amaranthus* spp. (Plansman – *A. hypochondriacus*, Pribina – *A. cruentus*, Zobor – *A. hypochondriacus x A. hybridus*), from which DNA was extracted from grains utilizing the commercial EliGene® Plant DNA Isolation Kit (Elisabeth Pharmacon). PCR products were separated on 1.5% agarose gels stained with GelRed® Nucleic Acid Gel Stain (Biotium) and visualized using a BDA digital system 30 transilluminator (Analytik Jena). As a control, DNA from wheat (*Triticum aestivum* L.) grain was used.

BLAST analyses revealed a single match between the genomes of the examined pseudocereals and wheat allergens, specifically to the Tri a 37 DNA sequence. This sequence corresponded to a predicted mRNA sequence of *Amaranthus tricolor* (see **Table 4**). As for the PCR analyses, neither of the studied allergens was confirmed in the examined samples (see **Figure 2**), which means that both studied pseudocereals, from the genetic point of view, can be considered safe for patients with wheat allergies; further tests are, of course, necessary.

Compared gene	Description	Scientific name	Query cover	E value	Per. identity	Accession
X96446.1	PREDICTED: Amaranthus tricolor serpin-ZX-like (LOC1,3081,0090), mRNA	*Amaranthus tricolor*	31%	2,00E-34	70.36%	XM_057676032.1

Table 4.
Results of BLAST analyses comparing the Tri a 37 wheat allergen to the genomes of Fagopyrum *spp. and* Amaranthus *spp.*

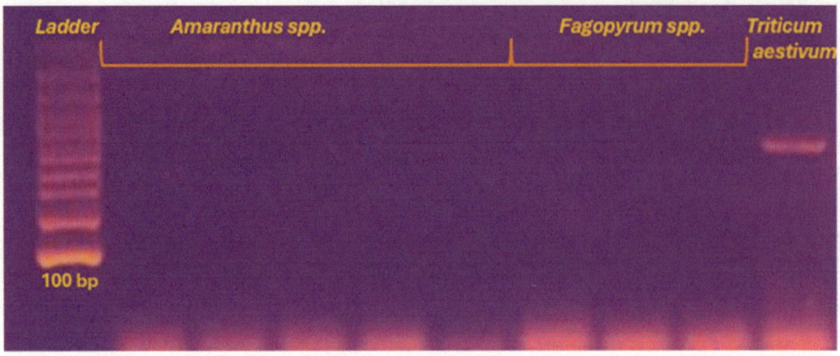

Figure 2.
Electrophoretogram of PCR analysis with primers for Tri a 37 wheat allergen homologs.

Our results correspond to the findings documented by others [68]. Zhao et al. [68] evaluated allergic cross-reactivity across dietary grains, utilizing SDS-PAGE, western blotting, indirect ELISA, and inhibitory ELISA, and evaluated serum from 12 individuals with wheat allergies. Insignificant cross-reactivity with buckwheat was noted.

5.2 Comparative analysis of *Fagopyrum* spp. allergen genes to the genomes of commonly consumed cereals and PCR validation

Seven allergens for buckwheat are identified, of which one has characterized epitopes for IgE, specifically the 13S globulin known as Fag e 1. A BLAST comparison indicated no significant matches when mRNA sequences were compared; however, for the amino acid sequence, the algorithm blastp identified similarity with 100 sequences (see **Figure 3**) in the NCBI database. The identified sequences include the 11 S globulin from amaranth (CAA57633.1), the 12 S globulin from oat (CAA54152.1), and the glutenin from barley (KAE8814294.1), as well as proteins from both spring wheat and durum wheat.

From the information obtained through the BLAST analysis, degenerate primers were designed to detect conserved gene regions of Fag e 1 using PCR methodology. The sequence of the degenerate forward primer was as follows: 5′–3′: GTSCTGGARACCCGYRYBGA, and for the reverse primer: 5′–3′: GCTAAAAATCGGGTTCGGGG.

The temperature profile for the PCR analysis of homolog detection was as follows: initial denaturation at 95°C, followed by 35 cycles comprising denaturation at 95°C for 15 seconds, primer annealing at 56°C for 15 seconds, and polymerization at 72°C for 45 seconds. Individual reactions were performed in a volume of 10 μL, utilizing MasterMix Robust HS Elizyme 2x (Elizabeth Pharmacon) with 400 nM of each primer. The DNA concentration for each reaction was set at 60 ng/μL. The biological material consisted of multiple cultivars of *Fagopyrum esculentum*, *Amaranthus* spp., *Triticum aestivum* L. (spring wheat), *Triticum durum* Desf. (durum wheat), *Avena sativa* L. (oats), and *Hordeum vulgare* L. (barley). DNA extraction and electrophoresis were conducted as mentioned above.

The primer pair produced polymorphic profiles across all crops. Notably, the fragment observed in all investigated crops, measuring 707 bp (see **Figure 4**), is an

DOI: http://dx.doi.org/10.5772/intechopen.1013800

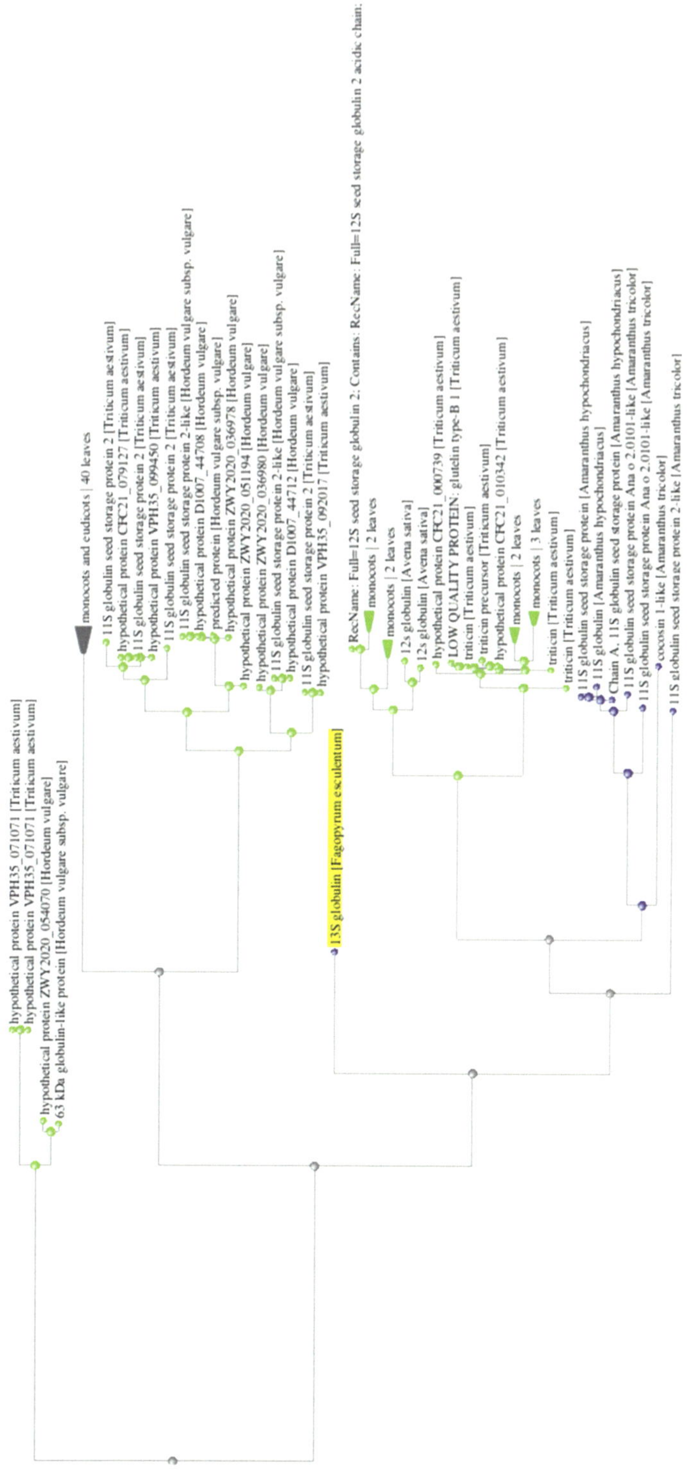

Figure 3.
Distance tree produced using BLAST pairwise alignments (NCBI, 2025).

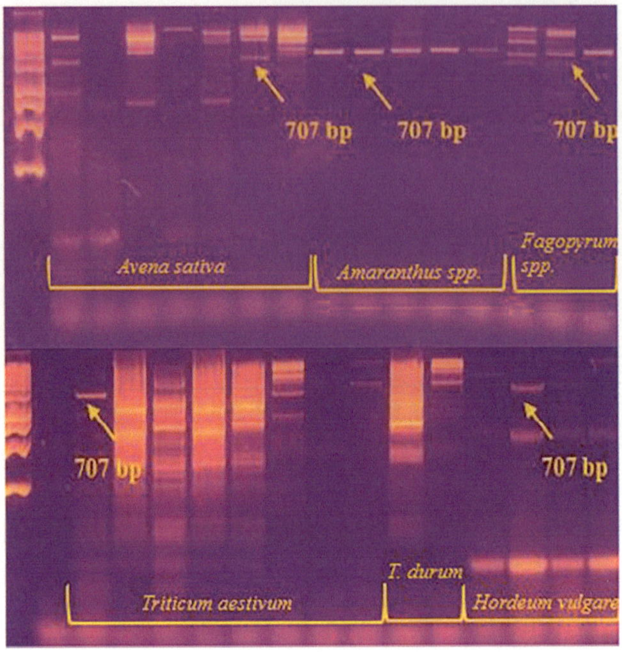

Figure 4.
Electrophoretogram of PCR screening of Fag e 1 homologs in Avena sativa *L.,* Amaranthus *spp.,* Fagopyrum *spp.,* Triticum aestivum *L.,* Triticum durum *Desf.,* Hordeum vulgare *L. genomes.*

interesting result, indicating that there may be sequence similarities between the allergen Fag e 1 and DNA sequences in wheat, oats, amaranth, and barley.

Buckwheat 13 S (Fag e 1) protein, as an allergen, has been the subject of numerous studies, several of which have focused on IgE epitope analysis and protein sequence analysis. It has been reported in the study by Yoshioka et al. [30] that the major epitopes of Fag e 1 have no homology to critical epitopes identified in allergenic proteins of other species, suggesting that they may be specific to buckwheat. However, Sordet et al. [69] reported that the IgE epitopes match, sequence-wise, IgE-binding epitopes previously characterized in soybean 11 S globulins.

5.3 Screening of pollen allergen homologs in the genomes of *Amaranthus* spp. and *Fagopyrum* spp.

Pollen allergen homologs were detected using primers designed for conserved regions of pollen profilin genes and the BBAP technique. Primers for the profilin gene region (NCBI accession number: AY792608.1) were utilized as a polymorphic marker technique, as demonstrated by Žiarovská and Zeleňáková [70]. Their findings indicated that the coding region of profilins can serve as a DNA-based molecular marker and is universally applicable to various plant species. Additionally, primers for the BBAP methodology were designed by Žiarovská and Urbanová [64] to elucidate genomic variability in length polymorphism, drawing on sequences of homologous genes related to the Bet v 1 group. The temperature profile for the profilin and BBAP methodology followed these steps: initial denaturation at 95°C, followed by 40 cycles of denaturation at 95°C for 45 seconds, primer annealing at 54°C for 45 seconds, and polymerization at 72°C for 35 seconds; the final polymerization was held at 72°C for 10 minutes [64]. Individual reactions were performed in a volume of 10 μL, utilizing MasterMix Robust HS Elizyme 2x (Elizabeth Pharmacon) with 400 nM of each primer. The DNA concentration for each reaction was set at 40 ng/μL. Primer sequences for profilin homolog detection were as follows: forward (5′–3′): ACCGGCCAA GATCTGGTTTT; reverse (5′–3′): AGGTAGTCTCCCAACCTCTCC. Biological material and DNA isolation followed the methodology mentioned above. PCR products were separated on 1.5% agarose gels.

The results of PCR analyses confirmed the presence of Bet v 1 and profilin homologs in the genomes of buckwheat and amaranth (see **Figure 5**). The study investigated different varieties of these crops, each displaying distinct profiles developed through the specified marker techniques.

As for profilins, the buckwheat varieties created a total of 20 fragments in the size range from 75 bp to 1,641 bp. The Špačínska 1 variety created fragments that were 75, 84, 632, and 900 bp in size. The Jana C1 variety produced fragments of 84, 146, 174, 274, 579, 900, and 1,641 bp. The Pyra variety yielded fragments that were 84, 656, 900, and 1,641 bp, while the Aďa variety yielded fragments that were 84, 328, 656, and 1,641 bp. As we can see, the sizes of the amplicons differed between the varieties, but the PCR product of 900 bp was present in all the studied buckwheat varieties. Regarding the varieties of amaranth, the three examined varieties collectively produced 10 PCR amplicons, with sizes ranging from 40 to 1,641 bp. The Plainsman variety yielded fragments measuring 531, 652, and 1,670 bp. In

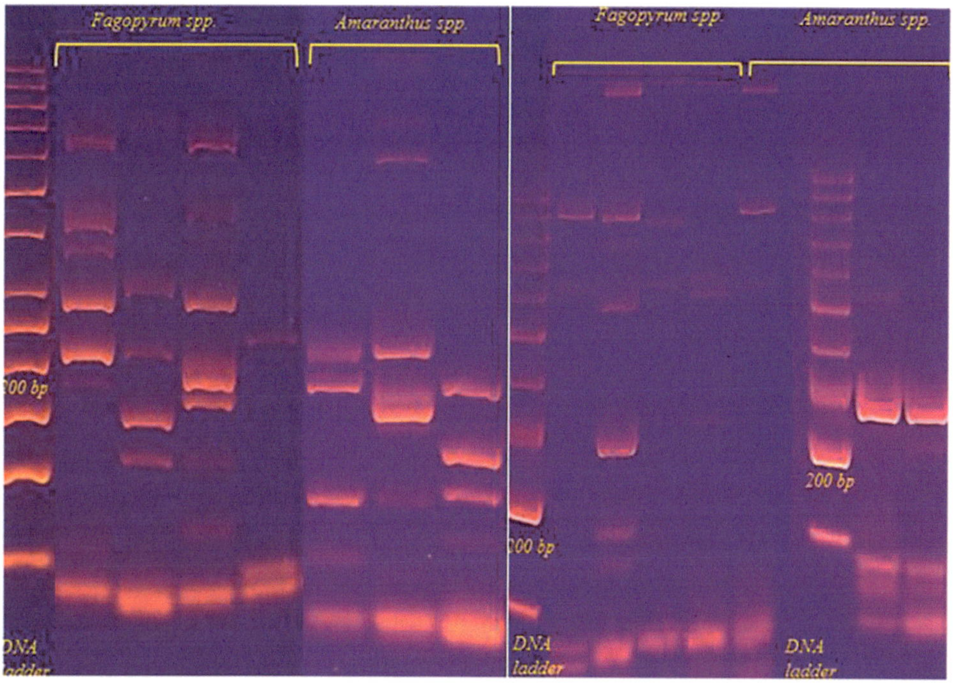

Figure 5.
Electrophoretogram of PCR screening of Bet v 1 (left) and profilin homologs (right) in the genomes of Fagopyrum *spp. and* Amaranthus *spp.*

contrast, the Pribina and Zobor varieties presented a similar profile, comprising amplicons of sizes 40, 84, and 263 bp. Notably, the Pribina variety also exhibited an additional fragment of 531 bp.

The BBAP methodology also effectively amplified Bet v 1 homologous DNA sequences in all studied varieties of amaranth and buckwheat. Amplification profiles were also polymorphic, similar to the case of profilins. For amplification, a degenerate primer pair was used according to the methodology of Žiarovská and Urbanová [64]. Buckwheat varieties generated amplicons ranging in size from 78 to 668 bp. The Špačínska 1 variety generated the most amplicons, with sizes of 170, 200, 260, 350, 418, and 668 bp. The Jana C1 variety produced amplicons with sizes of 114, 200, and 300 bp. The Pyra variety generated amplicons of sizes 668, 445, 260, 170, 114, and 80 bp. The Aďa variety had only one amplicon of size 200 bp. Amaranth varieties generated PCR products ranging from 80 to 1,026 bp. The Plansman variety exhibited a profile composed of amplicons measuring 124, 248, 280, and 330 bp. The Pribina variety yielded amplicons of sizes 124, 190, 330, 883, and 1,026 bp, while the Zobor variety generated amplicons of sizes 80, 124, 150, and 280 bp.

Owing to their significant sequence conservation, profilins constitute a family of cross-reactive allergens present in the pollens of both monocot and dicot plants, in addition to various plant-based foods [67]. Profilins, as panallergens, are associated with pollen-food allergy syndrome, wherein sensitization to pollen profilin may result in cross-reactive allergic reactions to fruits and vegetables containing profilin, including melon, tomato, and banana [71, 72]. While profilins are often regarded as minor allergens in terms of frequency, their clinical significance is amplified in polysensitized people and in areas with elevated pollen exposure [73].

Bet v 1, the principal allergen of birch (*Betula verrucosa*) pollen, is classified within the pathogenesis-related protein family 10 (PR-10) and significantly contributes to IgE-mediated sensitization in areas endemic to birch [74]. Bet v 1 demonstrates significant cross-reactivity with homologous PR-10 proteins present in various plant-based foods, such as apples (Mal d 1), hazelnuts (Cor a 1), celery (Api g 1), and soybeans (Gly m 4), owing to its conserved three-dimensional structure and sequence homology [75, 76].

6. Conclusion

This chapter summarized the currently known allergens of the pseudocereals, buckwheat, and amaranth. Amaranth is often regarded as a hypoallergenic crop, and to this date, there are no specific food allergens reported. Nonetheless, it is important to note that, as a pollen allergen, amaranth can pose challenges for individuals with heightened sensitivities. The buckwheat protein Fag e 1 is recognized as an allergen that can lead to severe reactions in allergic individuals, including anaphylaxis. Cross-allergies represent an increasingly concerning issue; the World Health Organization predicts that, by 2050, one in two people worldwide will experience some form of allergy [77]. Consequently, studying allergens at the genomic level is crucial. This approach facilitates cost-effective and rapid screening of homologous sequences, with findings that can be applied in subsequent research at the protein and immunology levels.

Acknowledgments

This publication was supported by the KEGA 001SPU-4/2025 – Internationalization of textbooks with a genetic and molecular-security background in the context of activating education, and the Operational Program Integrated Infrastructure within the project: Demand-driven research for sustainable and innovative food, Drive4SIFood 313011V336, cofinanced by the European Regional Development Fund.

Conflict of Interest

The authors declare no conflict of interest.

Author contribution

S. Farkasová (50 %): Writing, methodology, validation, investigation; J. Žiarovská (40%): Writing, methodology, supervision; E. Ivanišová (10%): Editing.

Author details

Silvia Farkasová[1], Jana Žiarovská[2] and Eva Ivanišová[3]*

1 Agrobiotech Research Centre, Slovak University of Agriculture in Nitra, Nitra, Slovak Republic

2 Institute of Plant and Environmental Sciences, Faculty of Agrobiology and Food Resources, Slovak University of Agriculture in Nitra, Nitra, Slovak Republic

3 Institute of Food Sciences, Faculty of Biotechnology and Food Sciences, Slovak University of Agriculture in Nitra, Chrenová, Slovak Republic

*Address all correspondence to: eva.ivanisova@uniag.sk

References

[1] Gorinstein S, Vargas OJM, Jaramillo NO, Salas IA, Ayala ALM, Arancibia-Avila P, Toledo F, Katrich E, Trakhtenberg S, et al. The total polyphenols and the antioxidant potentials of some selected cereals and pseudocereals. European Food Research and Technology. 2007;**225**(3–4):321–328. DOI: 10.1007/s00217-006-0417-7

[2] Gondola I, Papp P. Origin, geographical distribution and phylogenic relationships of common buckwheat (Fagopyrum esculentum Moench.). European Journal of Plant Science and Biotechnology. 2010;**4**(1):17–32

[3] Matsui K, et al. Genetic and genomic research for the development of an efficient breeding system in heterostylous self-incompatible common buckwheat (Fagopyrum esculentum). Theoretical and Applied Genetics. 2020;**133**(5):1641–1653. DOI: 10.1007/s00122-020-03572-6

[4] Yao H, et al. Deep sequencing of the transcriptome reveals distinct flavonoid metabolism features of black tartary buckwheat (Fagopyrum tataricum Garetn.). Progress in Biophysics and Molecular Biology. 2017;**124**:49–60

[5] Ohnishi O. On the origin of cultivated buckwheat. Advances in Buckwheat Research. A Proceedings of the 9th International Symposium on Buckwheat, Prague, Czech Republic. 2004. p. 16–21

[6] Fang X, et al. Genome-wide mining, characterization, and development of microsatellite markers in Tartary buckwheat (Fagopyrum tataricum Garetn.). Euphytica. 2019;**215**(11):1–10. DOI: 10.1007/s10681-019-2502-6

[7] Erquicia SP, et al. Rhinoconjunctivitis and occupational asthma due to buckwheat flour allergy. Archivos de Bronconeumología. 2020;**56**(7):466–468. DOI: 10.1016/j.arbres.2020.02.001

[8] Joshi DC, et al. From zero to hero: The past, present and future of grain amaranth breeding. Theoretical and Applied Genetics. 2018;**131**(9):1807–1823. DOI: 10.1007/s00122-018-3138-y

[9] Kreft S, et al. Extraction of rutin from buckwheat (Fagopyrum esculentum Moench) seeds and determination by capillary electrophoresis. Journal of Agricultural and Food Chemistry. 1999;**47**(11):4649–4652. DOI: 10.1016/j.foodchem.2005.05.081

[10] Steadman KJ, et al. Minerals, phytic acid, tannin and rutin in buckwheat seed milling fractions. Journal of the Science of Food and Agriculture. 2001;**81**(11):1094–1100. DOI: 10.1002/jsfa.914

[11] Luthar Z, et al. Tartary buckwheat in human nutrition. Plants. 2021;**10**(4):700. DOI: 10.3390/plants10040700

[12] Chettry U, Chrungoo N. Beyond the cereal box: Breeding buckwheat as a strategic crop for human nutrition. Plant Foods for Human Nutrition. 2021;**76**(4):399–409. DOI: 10.1007/s11130-021-00930-7

[13] Clouse JW, et al. The amaranth genome: Genome, transcriptome, and physical map assembly. The Plant Genome. 2016;**9**(1):2015.07.0062. DOI: 10.3835/plantgenome2015.07.0062

[14] Bressani R, et al. Chemical composition of grain amaranth cultivars

and effects of processing on their nutritional quality. Food Reviews International. 1992;**8**(1):23–49. DOI: 10.1080/87559129209540928

[15] Bressani R. Amaranth. In Encyclopedia of Food Sciences and Nutrition. 2nd. New York: Academic Press; 2003. p. 166–173. DOI: 10.1016/B0-12-227055-X/00036-5

[16] Ma X, et al. A chromosome-level Amaranthus cruentus genome assembly highlights gene family evolution and biosynthetic gene clusters that may underpin the nutritional value of this traditional crop. The Plant Journal. 2021;DOI: 10.1111/tpj.15298

[17] Pomés A, et al. WHO/IUIS allergen nomenclature: Providing a common language. Molecular Immunology. 2018;**100**:3–13

[18] Huby RDJ. Why are some proteins allergens? Toxicological Sciences. 2000;**55**(2):235–246. DOI: 10.1093/toxsci/55.2.235

[19] Dalal I, Binson I, Reifen R, Amitai Z, Shohat T, Rahmani S, et al. Food allergy is a matter of geography after all: Sesame as a major cause of severe IgE-mediated food allergic reactions among infants and young children in Israel. Allergy. 2002;**57**(4):362–365. DOI: 10.1034/j.1398-9995.2002.1s3412.x

[20] Gupta RS, Springston EE, Warrier MR, Smith B, Kumar R, Pongracic J, et al. The prevalence, severity, and distribution of childhood food allergy in the United States. Pediatrics. 2011;**128**(1):e9–17. DOI: 10.1542/peds.2011-0204

[21] Greenhawt M, Weiss C, Conte ML, Doucet M, Engler A, Camargo CA. Racial and ethnic disparity in food allergy in the United States: A systematic review. The Journal of Allergy and Clinical Immunology: In Practice. 2013;**1**(4):378–386. DOI: 10.1016/j.jaip.2013.04.009

[22] Gupta RS, Springston EE, Smith B, Warrier MR, Pongracic J, Holl JL. Geographic variability of childhood food allergy in the United States. Clinical Pediatrics. 2012;**51**(9):856–861. DOI: 10.1177/0009922812448526

[23] Koplin JJ, Peters RL, Ponsonby AL, Gurrin LC, Hill D, Tang MLK, et al. Increased risk of peanut allergy in infants of Asian-born parents compared to those of Australian-born parents. Allergy. 2014;**69**(12):1639–1647. DOI: 10.1111/all.1248

[24] Ebisawa M, et al. Food Allergy: Molecular Basis and Clinical Practice. Basel: Karger Medical and Scientific Publishers; 2015. DOI: 10.1159/000371647

[25] Valenta R, Hochwallner H, Linhart B, Pahr S. Food allergies: The basics. Gastroenterology. 2015;**148**(6): 1120–1131

[26] Bischoff S, Crowe SE. Gastrointestinal food allergy: New insights into pathophysiology and clinical perspectives. Gastroenterology. 2005;**128**(4):1089–1113

[27] Lockey RF, et al. Allergens and Allergen Immunotherapy. 3rd ed. New York: Marcel Dekker; 2004

[28] Larsen JN, et al. Epitope structure of recombinant isoallergens of Bet v 1. In: New Horizons in Allergy Immunotherapy. Berlin: Springer; 1996. p. 243–249. DOI: 10.1007/978-1-4615-5855-2_34

[29] Sicherer SH. Clinical implications of cross-reactive food allergens. The

Journal of Allergy and Clinical Immunology. 2001;**108**(6):881–890. DOI: 10.1067/mai.2001.118515

[30] Yoshioka H, et al. Expression and epitope analysis of the major allergenic protein Fag e 1 from buckwheat. Journal of Plant Physiology. 2004;**161**(7):761–767. DOI: 10.1016/j.jplph.2004.01.010

[31] Norbäck D, Wieslander G. A review on epidemiological and clinical studies on buckwheat allergy. Plants. 2021;**10**(3):607. DOI: 10.3390/plants10030607

[32] Sammut D, Dennison P, Venter C, Kurukulaaratchy RJ. Buckwheat allergy: A potential problem in 21st century Britain: Table 1. BMJ Case Reports. 2011;**2011**:bcr0920114882. DOI: 10.1136/bcr.09.2011.4882

[33] Badiu I, Olivieri E, Montagni M, Guida G, Mietta S, Pizzimenti S, et al. Italian study on buckwheat allergy: Prevalence and clinical features of buckwheat-sensitized patients in Italy. International Journal of Immunopathology and Pharmacology. 2013;**26**(3):801–806. DOI: 10.1177/039463201302600328

[34] Sordet C, et al. IgE-binding epitopic peptide mapping on a three-dimensional model built for the 13S globulin allergen of buckwheat (Fagopyrum esculentum). Peptides. 2009;**30**(6):1021–1027. DOI: 10.1016/j.peptides.2009.03.005

[35] Khan N, et al. Tandem repeat inserts in 13S globulin subunits, the major allergenic storage protein of common buckwheat (Fagopyrum esculentum Moench) seeds. Food Chemistry. 2012;**133**(1):29–37. DOI: 10.1016/j.foodchem.2011.12.056

[36] Milisavljević MD, et al. Vicilin-like storage globulin from buckwheat (Fagopyrum esculentum Moench) seeds. Journal of Agricultural and Food Chemistry. 2004;**52**(16):5258–5262. DOI: 10.1021/jf049519v

[37] Ayala-Niño A, et al. Amaranth proteins as a source of bioactive peptides: A review. International Food Research Journal. 2020;**27**(1). http://ifrj.upm.edu.my/27%20(01)%202020/01%20-%20IFRJ19167.R1.pd

[38] Siriwattanaakul U, et al. Amaranthus species around Bangkok, Thailand and the release of allergenic proteins from their pollens. Asian Pacific Journal of Allergy and Immunology 2015;**33**:203–210

[39] Kasera R, et al. First case report of anaphylaxis caused by Rajgira seed flour (Amaranthus paniculatus) from India: A clinico-immunologic evaluation. Asian Pacific Journal of Allergy and Immunology 2013;**31**:79–83

[40] Cárdenas-Torres FI, et al. Assessing the sensitizing and allergenic potential of the albumin and globulin fractions from amaranth (Amaranthus hypochondriacus) grains before and after an extrusion process. Medicina. 2019;**55**(3):72. DOI: 10.3390/medicina55030072

[41] Cressman RF, Ladics G. Further evaluation of the utility of "sliding window" FASTA in predicting cross-reactivity with allergenic proteins. Regulatory Toxicology and Pharmacology. 2009;**54**(3):S20–5. DOI: 10.1016/j.yrtph.2008.11.006

[42] Ladics GS, Bannon GA, Silvanovich A, Cressman RF. Comparison of conventional FASTA identity searches with the 80 amino acid sliding window

FASTA search for the elucidation of potential identities to known allergens. Molecular Nutrition & Food Research. 2007;**51**(8):985–998. DOI: 10.1002/mnfr.200600231

[43] Pearson WR, Lipman DJ. Improved tools for biological sequence comparison. Proceedings of the National Academy of Sciences. 1988;**85**(8):2444–2448. DOI: 10.1073/pnas.85.8.2444

[44] Brusic V, Petrovsky N. Immunoinformatics and its relevance to understanding human immune disease. Expert Review of Clinical Immunology. 2005;**1**(1):145–157. DOI: 10.1586/1744666X.1.1.145

[45] Tao A, Raz E. Allergy Bioinformatics. Dordrecht: Springer; 2015. DOI: 10.1007/978-94-017-7444-4

[46] Brusic V, Petrovsky N. Bioinformatics for characterisation of allergens, allergenicity and allergic cross-reactivity. Trends in Immunology. 2003;**24**(5):225–228. DOI: 10.1016/S1471-4906(03)00075-9

[47] Kadam K, Sawant S, Jayaraman VK, Kulkarni-Kale U. Databases and algorithms in allergen informatics. Bioinformatics - Updated Features and Applications. 2016. DOI: 10.5772/63083

[48] Monaci L, Visconti A. Immunochemical and DNA-based methods in food allergen analysis and quality assurance perspectives. Trends in Food Science & Technology. 2010;**21**(6):272–283. DOI: 10.1016/j.tifs.2010.02.003

[49] Sathe SK, Sharma GM. Effects of food processing on food allergens. Molecular Nutrition & Food Research. 2009;**53**(8):970–978. DOI: 10.1002/mnfr.200800194

[50] Sena-Torralba A, et al. Recent advances and challenges in food-borne allergen detection. Trends in Analytical Chemistry. 2020:116050. DOI: 10.1016/j.trac.2020.116050

[51] Zhang M, et al. Advanced DNA-based methods for the detection of peanut allergens in processed food. Trends in Analytical Chemistry. 2019;**114**:278–292

[52] Eischeid AC. Development and evaluation of a real-time PCR assay for detection of lobster, a crustacean shellfish allergen. Food Control. 2016;**59**:393–399. DOI: 10.1016/j.foodcont.2015.06.013

[53] Linacero R, et al. Detection by real time PCR of walnut allergen coding sequences in foods. Food Chemistry. 2016;**202**:334–340

[54] López-Calleja IM, et al. Development of a real time PCR assay for detection of allergenic trace amounts of peanut (Arachis hypogaea) in processed foods. Food Control. 2013;**30**(2):480–490. DOI: 10.1016/j.foodcont.2012.09.017

[55] Scaravelli E, et al. Development of three real-time PCR assays to detect peanut allergen residue in processed food products. European Food Research and Technology. 2008;**227**:857–869

[56] Sanchiz Á, et al. Detection of pistachio allergen coding sequences in food products: A comparison of two real time PCR approaches. Food Control. 2017;**75**:262–270

[57] López-Calleja IM, et al. Market analysis of food products for detection

of allergenic walnut (Juglans regia) and pecan (Carya illinoinensis) by real-time PCR. Food Chemistry. 2015; **177**:111–119

[58] Platteau C, et al. Detection of allergenic ingredients using real-time PCR: A case study on hazelnut (Corylus avellena) and soy (Glycine max). Journal of Agricultural and Food Chemistry. 2011;**59**(20):10803–10814. DOI: 10.1021/jf202110f

[59] Amiteye S. Basic concepts and methodologies of DNA marker systems in plant molecular breeding. Heliyon. 2021;**7**(10):e08093. DOI: 10.1016/j.heliyon.2021.e08093

[60] Collard BCY, Mackill DJ. Conserved DNA-derived polymorphism (CDDP): A simple and novel method for generating DNA markers in plants. Plant Molecular Biology Reporter. 2009;**27**(4):558–562. DOI: 10.1007/s11105-009-0118-z

[61] Yamanaka S, Suzuki E, Tanaka M, Takeda Y, JA W, KN W. Assessment of cytochrome P450 sequences offers a useful tool for determining genetic diversity in higher plant species. Theoretical and Applied Genetics. 2003;**108**(1):1–9. DOI: 10.1007/s00122-003-1403-0

[62] Poczai P, Varga I, Laos M, Cseh A, Bell N, Valkonen JP, et al. Advances in plant gene-targeted and functional markers: A review. Plant Methods. 2013;**9**(1):6. DOI: 10.1186/1746-4811-9-6

[63] Taberlet P, Gielly L, Pautou G, Bouvet J. Universal primers for amplification of three non-coding regions of chloroplast DNA. Plant Molecular Biology. 1991;**17**(5):1105–1109. DOI: 10.1007/bf00037152

[64] Žiarovská J, Urbanová L. Utilization of Bet v 1 homologs based amplified profile (BBAP) variability in allergenic plants fingerprinting. Biologia. 2022;**77**:517–523. DOI: 10.1007/s11756-021-00943-2

[65] Jimenez-Lopez JC, et al. Olive tree genetic background is a major cause of profilin (Ole e 2 allergen) polymorphism and functional and allergenic variability. Communications in Agricultural and Applied Biological Sciences 2013;**78**:213–219

[66] Radauer C, Breiteneder H. Evolutionary biology of plant food allergens. The Journal of Allergy and Clinical Immunology. 2007;**120**(3): 518–525. DOI: 10.1016/j.jaci.2007.07.024

[67] Jones SM, et al. Immunologic cross-reactivity among cereal grains and grasses in children with food hypersensitivity. The Journal of Allergy and Clinical Immunology 1995;**96**:341–351

[68] Zhao J, et al. Extraction of total wheat (Triticum aestivum) protein fractions and cross-reactivity of wheat allergens with other cereals. Food Chemistry. 2021;**347**:129064

[69] Sordet C, Culerrier R, Granier C, Didier A, Rougé P. IgE-binding epitopic peptide mapping on a three-dimensional model built for the 13S globulin allergen of buckwheat (Fagopyrum esculentum). Peptides. 2009;**30**(6):1021–1027. DOI: 10.1016/j.peptides.2009.03.005

[70] Žiarovská J, Zeleňáková L. Application of genomic data for PCR screening of Bet v 1 conserved sequence in clinically relevant plant species. Systems Biology. 2019. DOI: 10.5772/intechopen.80312

[71] Ebner C, et al. Identification of allergens in fruits and vegetables: IgE cross-reactivities with pollen allergens. The Journal of Allergy and Clinical Immunology. 1995;**95**(5 Pt 1):962–969. DOI: 10.1016/S0091-6749(95)70100-4

[72] Wensing M, et al. IgE to Bet v 1 and profilin: Cross-reactivity patterns and clinical relevance. The Journal of Allergy and Clinical Immunology. 2002;**110**(3):435–442. DOI: 10.1067/mai.2002.127714

[73] Hauser M, et al. Allergens from pollen and fruit: Cross-reactivity and allergen structure. Methods. 2010;**55**(4):297–307. DOI: 10.1016/j.ymeth.2011.01.005

[74] Vieths S, Scheurer S, Ballmer-Weber B. Current understanding of cross-reactivity of food allergens and pollen. Annals of the New York Academy of Sciences. 2002;**964**:47–68. DOI: 10.1111/j.1749-6632.2002.tb04134.x

[75] Focke M, et al. Non-anaphylactic food allergy: Bet v 1 as a possible link between birch pollen and hazelnut allergy. The Journal of Allergy and Clinical Immunology. 2003;**111**(2):427–432. DOI: 10.1067/mai.2003.121

[76] Mittag D, et al. Ara h 8, a Bet v 1-homologous protein from peanut, cross-reacts with Bet v 1 and correlates with birch pollen allergy. Clinical and Experimental Allergy. 2004;**34**(9):1426–1433. DOI: 10.1111/j.1365-2222.2004.02051.x

[77] Tanno LK, Chalmers R, Jacob R, Kostanjsek N, Bierrenbach AL, Martin B, et al. Global implementation of the world health organization's International Classification of Diseases (ICD)-11: The allergic and hypersensitivity conditions model. Allergy. 2020;**75**(9):2206–2218. DOI: 10.1111/all.14468

Section 2

Reimagining Cereal and Pseudocereal Systems for the 21st Century

Chapter 3

Reimagining Grains: Innovative Paths for Cereals and Pseudocereals

Pramod Kumar Sharma, Kadadanamari S. Shashidhar, S. Bhuvaneshwari, Mahipal Choubey and Ajeet Singh

Abstract

Cereals and pseudocereals are foundational to global food and nutritional security. While staple cereals like rice, wheat, and maize dominate agriculture, nutrient-dense pseudocereals – quinoa, amaranth, buckwheat, and chia – are emerging as vital alternatives. This chapter explores the evolving role of these grains in the twenty-first century, focusing on production systems, climate resilience, nutritional value, and market dynamics. Often termed "forgotten grains," pseudocereals are transforming agricultural and dietary paradigms by offering solutions where conventional cereals often fall short. Traditionally cultivated with minimal inputs on marginal lands, these crops are gaining renewed attention for their sustainability and hardiness. Quinoa, revered as the "mother grain" of the Incas; amaranth, once sacred to the Aztecs; buckwheat, a versatile non-cereal; and chia, the ancient energy seed – all exemplify this untapped potential. This analysis highlights how pseudocereals advance sustainable agriculture and nutrition, addressing contemporary challenges and future opportunities for the global food system.

Keywords: cereals, pseudocereals, food security, sustainable agriculture, climate resilience, quinoa, forgotten grains

1. Introduction

Cereals have been the foundation of human civilization for more than 10,000 years, serving as the primary source of sustenance and shaping agricultural, economic, and cultural systems worldwide [1, 2]. Crops such as rice (*Oryza sativa* L.), wheat (*Triticum aestivum* L.), and maize (*Zea mays* L.) remain indispensable staples, providing more than 60% of the global population's dietary energy needs [3]. They not only secure caloric intake but also underpin rural livelihoods, agricultural trade, and national economies [4]. The domestication of cereals marked the beginning of settled agriculture and facilitated the development of societies, making them central to food systems even in the twenty-first century [5].

However, the challenges of the modern era extend beyond mere production. While the Green Revolution of the mid-twentieth century dramatically boosted cereal yields and alleviated large-scale hunger, it also brought unintended consequences – soil degradation, groundwater depletion, loss of crop diversity, and rising dependence on chemical inputs [6, 7]. Moreover, cereals, though rich in carbohydrates, are relatively poor sources of certain essential amino acids, vitamins, and micronutrients such as iron and zinc [8]. This nutritional imbalance contributes to the so-called "hidden hunger" or micronutrient malnutrition, which affects over two billion people worldwide [9]. Thus, although cereals remain the backbone of food security, they alone cannot guarantee nutrition security in the present century.

In parallel, pseudocereals – a group of non-grass species that produce small, starchy seeds consumed much like cereals – have gained increasing recognition for their potential to complement and diversify modern diets [10, 11]. Unlike cereals, pseudocereals such as quinoa (*Chenopodium quinoa* Willd.), amaranth (*Amaranthus* spp.), and buckwheat (*Fagopyrum esculentum* Moench) are rich in high-quality proteins, essential amino acids, dietary fiber, minerals, and bioactive compounds [12]. Importantly, they are naturally gluten-free, positioning them as valuable alternatives for people with celiac disease and gluten intolerance [13], while also catering to the global health and wellness food market. Their adaptability to marginal soils, resilience under drought and salinity, and suitability for high-altitude farming make them highly relevant in the context of climate change and resource-constrained agriculture [11].

The twenty-first century poses a dual challenge: feeding a growing population while ensuring dietary quality and environmental sustainability. The world population is projected to reach 9.7 billion by 2050, demanding an estimated 60% increase in food production [14]. Yet, at the same time, agriculture is under mounting stress from climate variability, land and water scarcity, declining soil fertility, and biodiversity loss [15, 16]. The dominance of a few major cereals – rice, wheat, and maize – exacerbates vulnerability by narrowing genetic diversity and dietary options [17]. In this context, pseudocereals are increasingly being considered as "future smart crops" that can enhance nutritional diversity, strengthen climate resilience, and reduce over-reliance on a handful of staple grains.

The global food system, long reliant on a narrow base of staple cereals, stands at a pivotal juncture. While crops like rice (*Oryza sativa* L.), wheat (*Triticum aestivum* L.), and maize (*Zea mays* L.) will continue to be fundamental to caloric security, the twenty-first century demands a radical reimagining of our grain sources. This chapter posits that the path forward lies not in replacing these staples, but in innovatively integrating them with nutritionally superior and climate-resilient pseudocereals. The legacy of the Green Revolution, despite its monumental success in averting famine, has also led to ecological strain, genetic uniformity, and dietary imbalances, manifesting in the persistent challenge of hidden hunger. In this context, pseudocereals – such as quinoa (*Chenopodium quinoa* Willd.), amaranth (*Amaranthus* spp.), and buckwheat (*Fagopyrum esculentum* Moench) – emerge not merely as alternative crops, but as catalysts for a systemic transformation. They represent a convergence of ancient wisdom and modern scientific potential, offering unique nutritional profiles, functional benefits, and agronomic resilience. This chapter explores the innovative paths – across breeding, agronomy, market development, and policy – required to leverage the synergistic potential of both cereals and pseudocereals. It is a call to move beyond conventional paradigms and forge a more resilient, nutritious, and sustainable future for global agriculture.

2. Global production trends of cereals and pseudocereals

2.1 Overview of global production

Global cereal production has undergone a remarkable transformation over the past six decades. In 1961, total cereal production stood at approximately 877 million tons, constrained by traditional farming methods, limited irrigation, and low adoption of chemical fertilizers [18]. Food insecurity was widespread, particularly in Asia and Africa, and the focus was primarily on expanding cultivated areas rather than enhancing productivity per hectare.

By 1970, production had increased to 1,150 million tons, reflecting the initial impacts of the Green Revolution. The introduction of high-yielding wheat and rice varieties, improved irrigation infrastructure, chemical fertilizers, and mechanization resulted in a roughly 31% increase in production within a decade [4, 5]. During the 1980s, production accelerated further to 1,600 million tons, driven by the adoption of improved varieties, especially in Asia and Latin America. Maize gained prominence in the Americas, while rice and wheat continued to dominate Asian cereal production.

The 1990s saw a slight slowdown in growth, with production reaching 1,800 million tons, reflecting diminishing yield gains in intensively cultivated regions and limited scope for the expansion of arable land [6]. By the early 2000s, production increased to 2,100 million tons due to hybrid maize, improved wheat varieties, and expanded irrigation infrastructure. Climate-resilient varieties and biofortification programs also started emerging during this period [7].

By the 2010s, global cereal production climbed to 2,400 million tons, supported by precision agriculture, improved seed technologies, and optimized fertilizer management. During this decade, minor cereals and pseudocereals, including millets, quinoa, and amaranth, began receiving increasing attention for their nutritional value and resilience on marginal lands [9, 10]. Finally, between 2020 and 2022, global production reached a record 2,800 million tons. This reflects decades of cumulative technological progress and improved agronomic practices, but the plateauing of production also highlights emerging challenges such as extreme climate events, water scarcity, and soil degradation, particularly in South Asia, Sub-Saharan Africa, and parts of Latin America [16].

2.2 Major cereals versus pseudocereals

Cereals continue to dominate global grain production, with maize, wheat, and rice together accounting for approximately 91% of total cereal output in 2023 [19] (see **Figures 1a–c**). Other cereals, including barley, oats, sorghum, and minor millets, occupy a relatively small share of the global market. In contrast, pseudocereals such as quinoa, amaranth, and buckwheat contribute only a tiny fraction of total grain production (see **Figure 1b**). For example, global quinoa production is concentrated in Peru (~100–120 thousand tons) and Bolivia (tens of

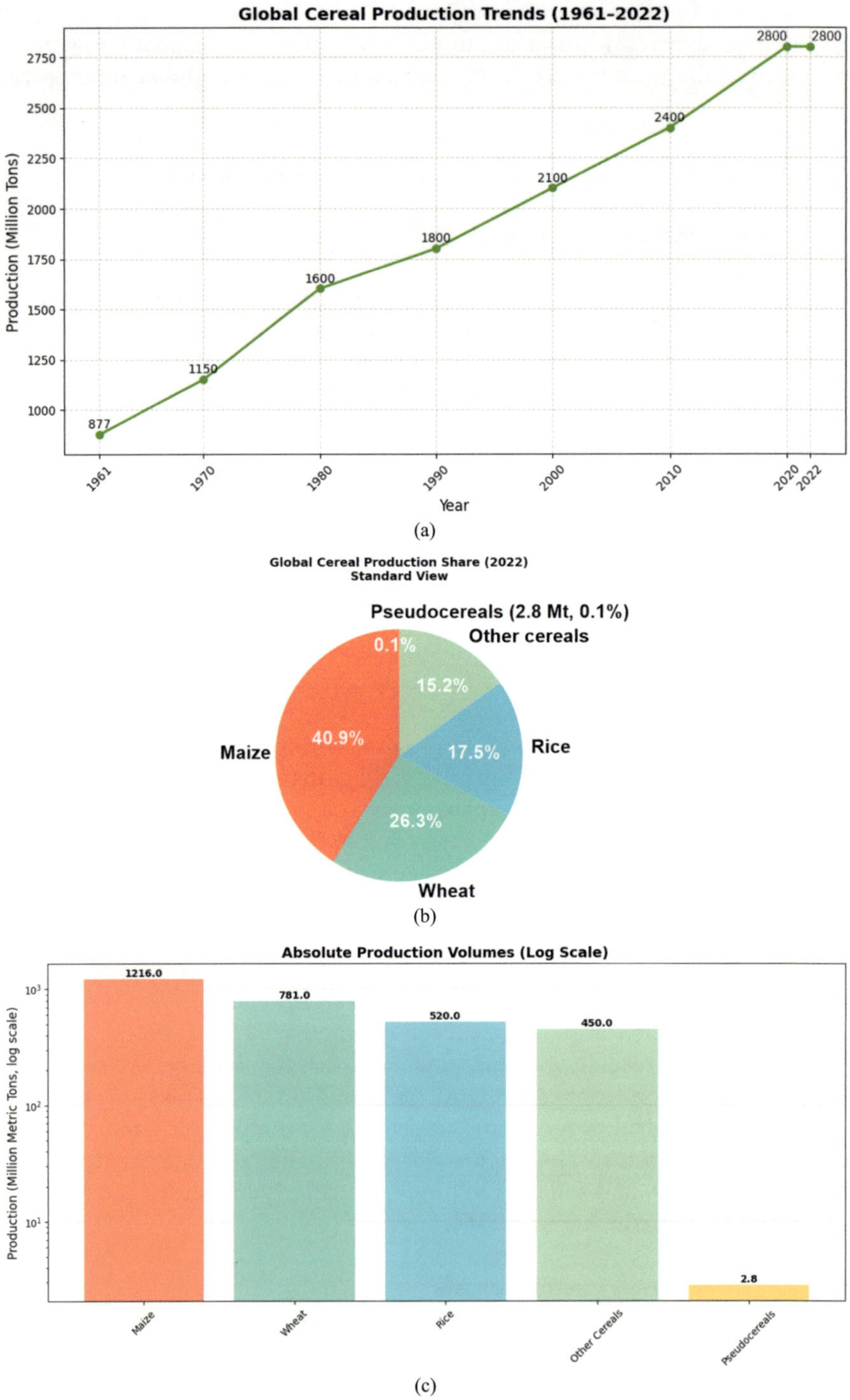

Figure 1.
Global production share of major cereals versus pseudocereals (2022). (a) Global cereal production trends (1961 to 2022). (b) Global production trends of cereals. (c) Absolute production volumes of cereals and pseudocereals. (The figures mentioned above are a visual representation showing the dominance of maize, wheat, and rice compared to the combined total of all pseudocereals (quinoa, amaranth, buckwheat). Other cereals (barley, oats, sorghum, millets) are included as a small sliver. Data source: FAOSTAT, 2023).

thousands of tons) [10, 11]. Despite their small production volume, pseudocereals occupy high-value market niches due to their nutritional quality, gluten-free properties, and increasing demand from health-conscious consumers.

2.3 Decade-wise cereal production trends

Global cereal production has increased markedly from "877 million tons in 1961" to approximately "2,800 million tons by 2022," reflecting sustained advancements in crop improvement, mechanization, irrigation, and input use. Early gains were largely driven by the Green Revolution, which introduced high-yielding wheat and rice varieties and significantly improved food availability, particularly in Asia.

From the 1980s onward, agricultural intensification and mechanization expanded across developing regions, leading to rapid productivity gains. However, by the 1990s, yield growth began to plateau in highly productive systems due to diminishing returns on inputs and increasing pressure on soil and water resources. Subsequent production increases were achieved through technological diversification, improved hybrids, and better crop management rather than area expansion.

The 2010s marked a transition toward "precision agriculture," optimized nutrient management, and data-driven decision-making, contributing to further increases in cereal output. Despite these advances, global cereal production has shown signs of stabilization since 2020, largely due to "climate-related stresses," including droughts, floods, and heatwaves. These trends underscore the vulnerability of cereal-dominated food systems to climate variability and change.

2.3.1 Pseudocereals, climate resilience, and sustainable development goals

Pseudocereals such as quinoa, amaranth, and buckwheat have emerged as promising crops for addressing the dual challenges of climate change and malnutrition. These crops exhibit high tolerance to abiotic stresses, including drought, salinity, and temperature extremes, and are well-suited to marginal and rainfed agro-ecosystems (see **Figures 2–4**).

Nutritionally, pseudocereals provide high-quality proteins with balanced amino acid profiles, essential minerals, dietary fiber, and bioactive compounds. Their incorporation into food systems directly supports "SDG 2 (Zero Hunger)" by enhancing food and nutritional security, and "SDG 3 (Good Health and Well-being)" by addressing micronutrient deficiencies and diet-related health issues.

From a climate perspective, diversification with pseudocereals aligns with "SDG 13 (Climate Action)" by promoting climate-resilient agriculture and reducing dependence on a narrow range of climate-sensitive cereals. Their inclusion in diversified cropping systems also supports "SDG 12 (Responsible Consumption and Production)" through sustainable resource use and resilient supply chains.

The stabilization of global cereal production under increasing climatic uncertainty highlights the need for "policy-driven diversification" beyond major cereals. Integrating pseudocereals into national food and nutrition security strategies aligns with climate policies, such as "climate-smart agriculture (CSA)" frameworks, "Nationally Determined Contributions (NDCs)" under the Paris Agreement, and sustainable agri-food system transitions.

Future food security will depend on synergizing productivity enhancement in major cereals with diversification toward resilient, nutrient-dense crops.

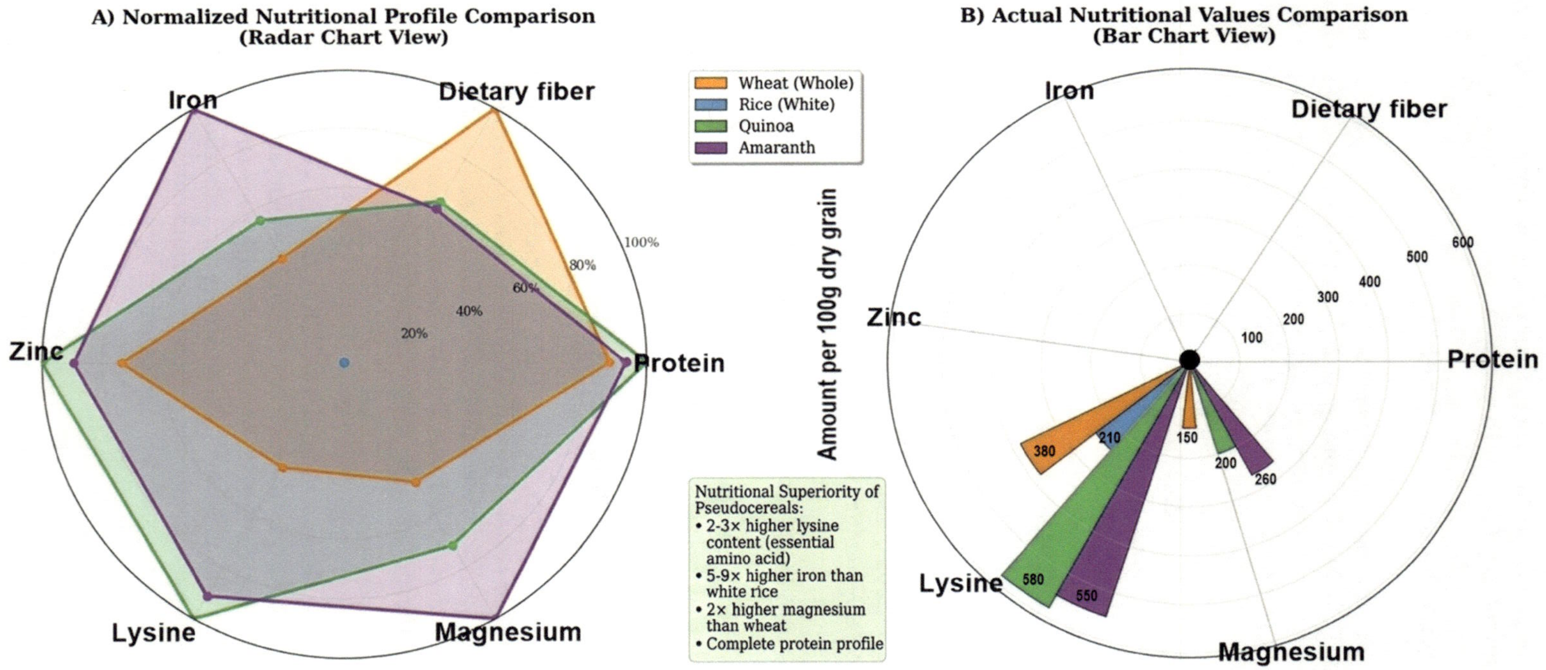

Figure 2.
Comparative nutritional profile of cereals and pseudocereals. (a) Normalized nutritional profile (radar chart): highlights the relative nutrient strengths of each crop. (b) Actual nutritional values (bar chart): compares energy, protein, fiber, and micronutrient content per 100 g. (Data sources: USDA FoodData Central, 2023; values per 100 g dry grain).

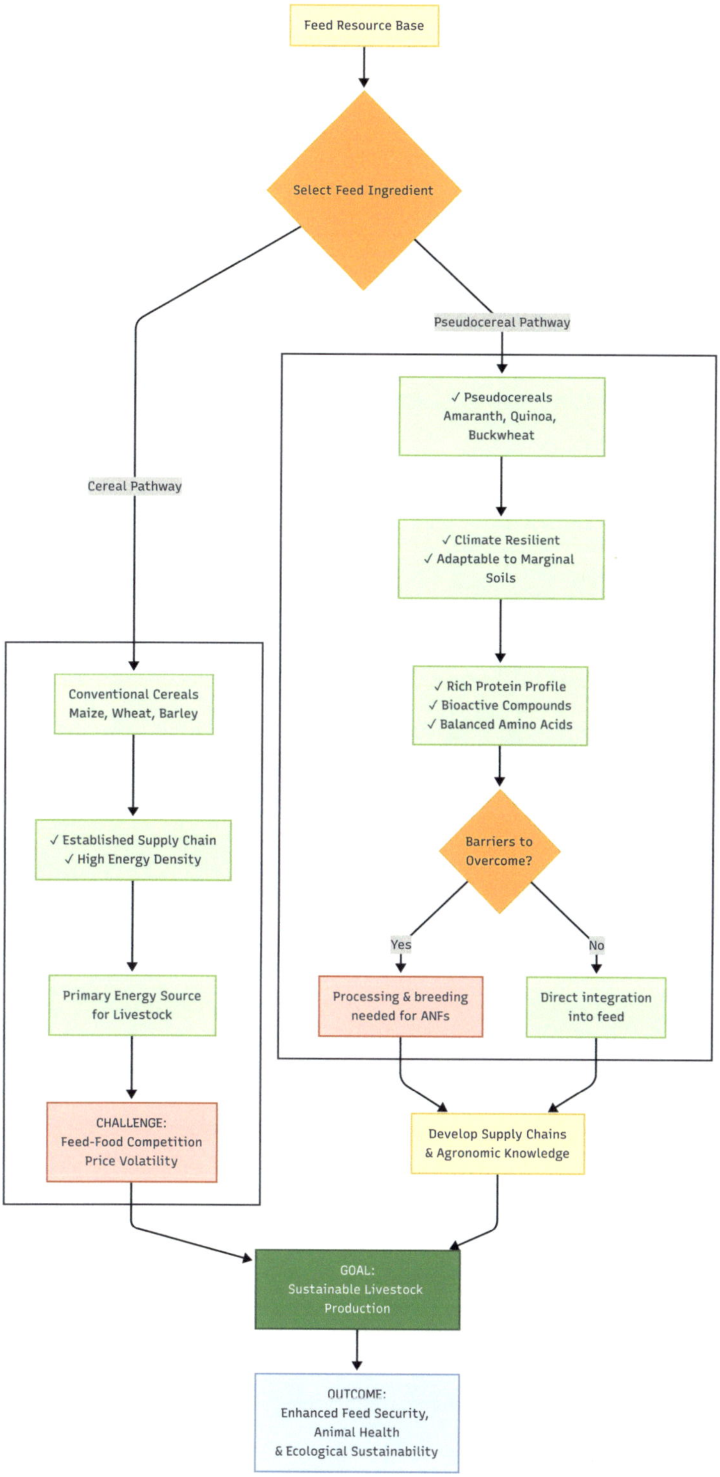

Figure 3.
Pathway to introduce cereals and pseudocereals into the animal feed system.

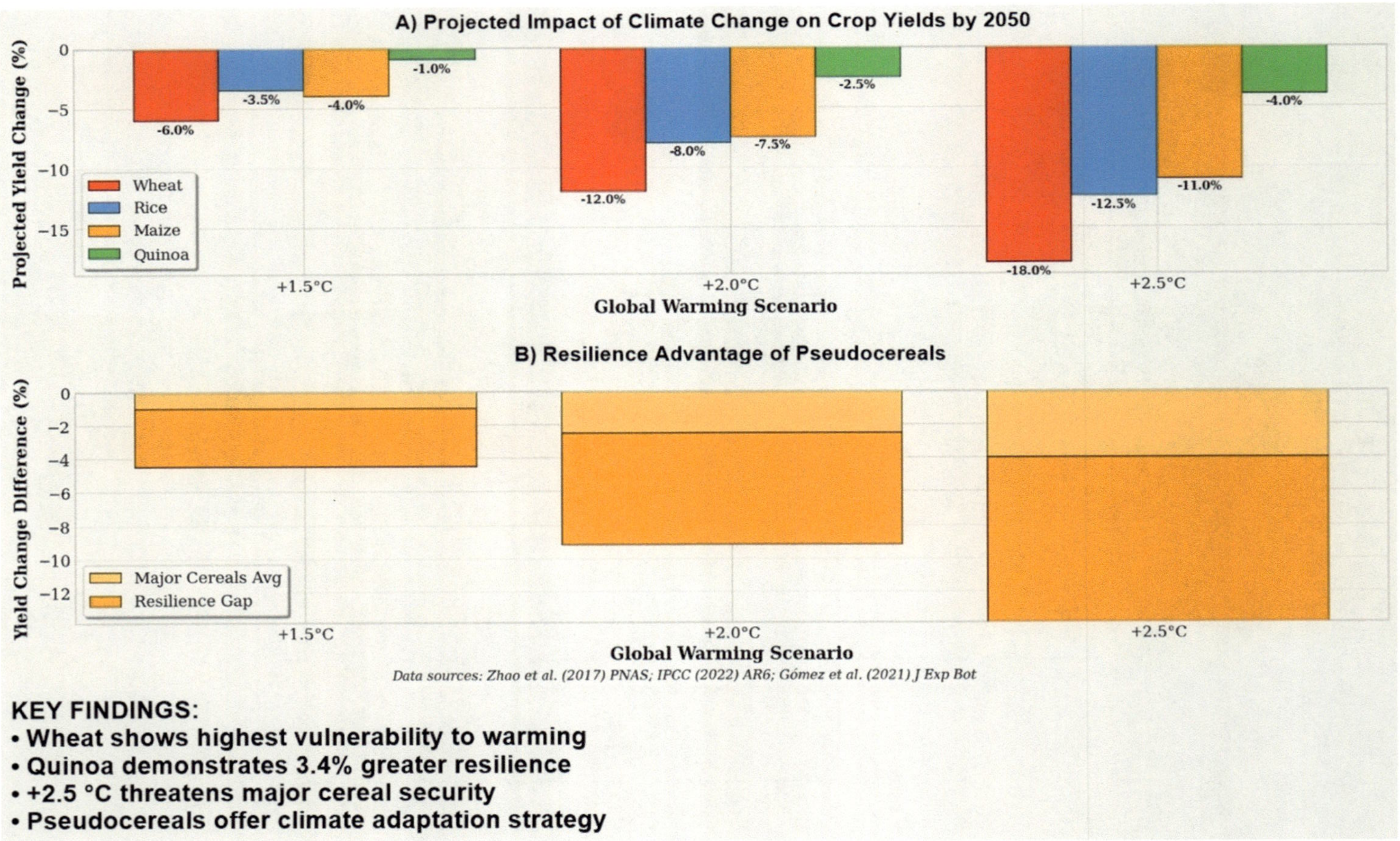

Figure 4.
Projected impact of climate change on cereal yield by 2050. A conceptual framework illustrating how integrating pseudocereals into mainstream agriculture addresses twenty-first-century challenges and creates synergistic opportunities for a more resilient and nutritious food future. Data Sources: Studies from literature [16, 24].

Mainstreaming pseudocereals through supportive policies, research investments, and market development can significantly contribute to sustainable, climate-resilient, and nutrition-sensitive food systems.

2.4 Implications

The global production trends reveal several important implications for future food system resilience and sustainability. The strong concentration of output in a few major cereals – maize, wheat, and rice – creates a structural vulnerability, as any disruption caused by climate extremes, pest outbreaks, or market fluctuations can have far-reaching consequences for global food security [17]. Although pseudocereals currently contribute relatively little to total production, their rising importance stems from their superior nutritional qualities and strong adaptability to climate stressors, making them valuable alternatives in diversified cropping systems (see **Table 1**). At the same time, global cereal markets remain highly sensitive to weather conditions, production estimates, and policy decisions, leading to periodic volatility in both trade flows and prices [18]. These dynamics underscore the need to diversify food systems by integrating pseudocereals more systematically into regional and global markets, especially in marginal environments and in segments where health and nutrition considerations drive consumer demand [17].

3. Nutritional dimensions: Addressing hidden hunger

While cereals remain the primary source of dietary energy globally, they are often deficient in essential amino acids, vitamins, and micronutrients. For example, rice is

Year	Global cereal production (million tons)	Key notes and trends
1961	877	Post-WWII era; low mechanization, limited irrigation, and fertilizer use constrained production.
1970	1,150	Early impacts of the Green Revolution; introduction of high-yielding wheat and rice varieties.
1980	1,600	Widespread adoption of improved varieties; mechanization expanded in Asia and Latin America.
1990	1,800	Yield gains plateau in highly productive regions; cultivated area expanded mainly in Africa and South America.
2000	2,100	Expansion of improved maize and wheat varieties; early adoption of biofortified and climate-resilient crops.
2010	2,400	Precision agriculture and optimized fertilizer use; pseudocereals gain attention for their nutritional and resilience traits.
2020	2,800	Record production due to climate-smart varieties and improved agronomy; plateauing in some regions due to climate stress.
2022	2,800	Production stabilizes; climate-related risks (droughts, floods, heatwaves) begin to affect vulnerable regions.

(Data source: FAO STAT, 2023)

Table 1.
Decade-wise cereal production trends.

Crop	Energy (kcal)	Protein (g)	Carbohydrates (g)	Fat (g)	Fiber (g)	Iron (mg)	Zinc (mg)	Lysine (g)
Wheat (whole grain)	340	13.2	72.6	2.5	12.2	3.6	2.7	0.30
Rice (polished)	365	7.1	80.4	0.7	1.3	0.8	1.2	0.09
Maize (yellow)	365	9.4	74.3	4.7	7.3	2.7	2.2	0.22
Quinoa	368	14.1	64.2	6.1	7.0	4.6	3.1	0.78
Amaranth	371	13.6	65.2	7.0	6.7	7.6	2.9	0.83
Buckwheat	343	13.3	71.5	3.4	10.0	2.2	2.4	0.68

Data sources: USDA, 2023.

Table 2.
Selected nutritional composition of cereals and pseudocereals (per 100 g edible portion).

notably low in lysine, maize is deficient in tryptophan and niacin, and wheat contains limited bioavailable iron and zinc (7, 2). The widespread reliance on refined cereals has contributed to a phenomenon termed "hidden hunger," wherein caloric sufficiency masks underlying micronutrient deficiencies, affecting over two billion people worldwide [8].

In contrast, pseudocereals – including quinoa (*Chenopodium quinoa* Willd.), amaranth (*Amaranthus* spp.), buckwheat (*Fagopyrum esculentum* Moench), and chia (*Salvia hispanica* L.) – offer superior nutritional profiles (see **Table 2**). Quinoa provides all nine essential amino acids, amaranth is rich in calcium and iron, and buckwheat is a significant source of rutin, a bioactive flavonoid with potent antioxidant properties [9, 10]. These properties make pseudocereals particularly valuable in combating malnutrition, supporting digestive health, and reducing risks associated with non-communicable diseases.

3.1 Nutritional comparison between cereals and pseudocereals

A nutritional comparison of cereals and pseudocereals reveals that, while their energy densities are comparable – ranging from approximately 340 to 370 kcal per 100 g, making pseudocereals viable staple foods – their nutritional profiles differ significantly (see **Table 2**). Pseudocereals such as quinoa, amaranth, and buckwheat offer distinct advantages, primarily through their superior protein quality. They not only contain higher protein levels but also possess a more balanced amino acid profile, being particularly rich in lysine, an essential amino acid often deficient in major cereals like wheat, rice, and maize. Furthermore, pseudocereals are more micronutrient-dense, with notably higher concentrations of essential minerals like iron and zinc, which can help address micronutrient deficiencies. They also contribute higher dietary fiber for improved digestive health and glycemic control and contain moderately higher levels of fat, which is predominantly composed of beneficial unsaturated fatty acids supportive of cardiovascular health. These combined attributes position pseudocereals as excellent candidates for crafting nutritionally enriched diets, gluten-free products, and functional foods aimed at combating malnutrition and lifestyle-related diseases.

3.2 Visual representation of pseudocereals

Pseudocereals such as quinoa, amaranth, and buckwheat exhibit superior nutritional quality compared to major cereals like wheat, rice, and maize. While energy content is broadly similar across cereals and pseudocereals (~340–370 kcal per 100 g), pseudocereals contain higher protein levels and more balanced amino acid profiles, particularly lysine, which is limited in most cereals (see **Figure 2a** and **b**). They are also richer in essential minerals such as iron and zinc and have higher dietary fiber content, contributing to better micronutrient intake and digestive health. Fat content is moderately higher in pseudocereals, predominantly comprising healthy unsaturated fatty acids. These characteristics make pseudocereals valuable for addressing protein-energy malnutrition and for use in gluten-free and functional food products.

4. Cereals and pseudocereals as animal feeds: Opportunities and challenges

Cereals such as maize, barley, wheat, sorghum, and oats have long been the backbone of animal feed systems worldwide, primarily due to their high carbohydrate content, energy density, and relative availability. They provide a consistent source of metabolizable energy, support efficient growth and production in livestock, and can be easily incorporated into feed rations. In recent decades, pseudocereals such as amaranth, quinoa, and buckwheat have attracted attention as alternative feed resources owing to their rich protein profile, balanced amino acid composition, and presence of bioactive compounds that may improve animal health and productivity [20]. Their adaptability to marginal environments and climate resilience also make them promising candidates to complement or partially replace conventional cereals in feed formulations, thereby contributing to sustainable livestock production systems.

Despite these opportunities, several challenges hinder the large-scale adoption of cereals and pseudocereals in animal nutrition (see **Figure 3**). Rising demand from the food industry increases competition for staple cereals, raising concerns over feed-food security [12]. Pseudocereals, though nutritionally superior, are often limited by anti-nutritional factors such as saponins and tannins, variability in nutrient availability (see **Table 3**), and higher production costs compared to traditional feed grains [21, 22]. Moreover, inconsistent supply chains and limited agronomic knowledge restrict their integration into mainstream feed markets. Addressing these challenges requires breeding strategies to reduce anti-nutritional compounds, development of cost-effective processing technologies, and promotion of diversified cropping systems that enhance both feed security and ecological sustainability [23]. Thus, cereals and pseudocereals together hold significant potential for the future of animal nutrition, provided research, policy, and market frameworks are aligned to overcome current barriers.

5. Climate change and resilience challenges

Climate change poses significant threats to global cereal production. Rising temperatures, altered rainfall patterns, and an increased frequency of extreme

Parameter	Cereals (maize, wheat, barley, sorghum, oats)	Pseudocereals (amaranth, quinoa, buckwheat)	Implications for animal feed
Energy (kcal/kg DM)	High (approx. 3,200–3,600 kcal/kg)	Moderate (approx. 3,000–3,400 kcal/kg)	Cereals provide consistent energy; pseudocereals supplement energy.
Protein (%)	Moderate (8%–12%)	High (12%–18%)	Pseudocereals improve protein intake and amino acid balance.
Amino acid profile	Limited in lysine and methionine	Well-balanced, complete essential amino acids	Enhances growth, reproduction, and health
Bioactive compounds	Low	Moderate–high (saponins, flavonoids)	Can improve immunity, antioxidant status
Anti-nutritional factors	Low	Saponins, tannins	May require processing or treatment before feeding.
Availability	Widely grown; stable supply	Limited production; niche crops	Cereals dominate feed markets; pseudocereals as supplements.
Cost	Low–moderate	Higher	Pseudocereals may increase feed cost
Environmental adaptation	Well-adapted to conventional conditions	Adaptable to marginal lands and climate stress.	Pseudocereals can diversify feed resources under climate change.

Sources: Studies by Zhaoa et al. [24] Jarvis et al. [25].

Table 3.
Comparative table – nutritional profile and feed potential.

weather events jeopardize both yields and crop stability. Wheat, for example, is highly sensitive to heat stress during anthesis, while rice requires substantial water inputs – resources that are increasingly limited in drought-prone regions. Maize faces compounded risks from both drought and pest outbreaks, highlighting the vulnerability of major cereals under changing climatic conditions (see **Figure 4**).

In contrast, pseudocereals demonstrate notable resilience in marginal and stress-prone environments. Quinoa thrives under saline and drought conditions, amaranth adapts well to degraded soils, and buckwheat matures rapidly within 90–100 days, making it ideal for short growing seasons. These adaptive traits position pseudocereals as strategic crops for building climate-resilient food systems.

5.1 Conceptual framework for an integrated cereal–pseudocereal system

To address nutritional security, climate resilience, and sustainability, an integrated food system approach is essential. The conceptual framework (see **Figure 4**) links key challenges (climate stress, nutritional deficiencies) to strategic opportunities (agricultural diversification, policy reform, biofortification), ultimately supporting the overarching goal of a sustainable and nutritious food system.

6. Market development and consumer trends

Global cereal markets remain robust due to staple food demand and industrial uses such as feed, starch, and ethanol. However, the growing health-conscious

Product category	Key drivers	Projected CAGR* (2023–2030)	Example products	Target consumer segment
Gluten-free bakery	Rising celiac/gluten sensitivity; health trends	~9.5%	Quinoa bread, buckwheat muffins, amaranth crackers	Health-conscious, gluten-free consumers
Plant-based proteins	Sustainability concerns; vegan/vegetarian diets	~10.5%	Quinoa protein powder, meat analogs with amaranth	Flexitarians, athletes, ethical consumers
Ancient grains/ superfoods	Demand for nutrient-dense, "clean-label" ingredients	~8.0%	Ancient grain breakfast cereals, quinoa salad blends	Wellness-focused, premium market
Gluten-free baby food	Parental focus on early-life nutrition and allergies	~7.5%	Quinoa-based infant cereals, teething biscuits	Parents of infants and toddlers
Non-dairy beverages	Lactose intolerance; dairy alternatives trend	~12.0%	Amaranth milk, quinoa drink	Lactose-intolerant, vegan consumers

**CAGR: compound annual growth rate (estimates are based on market research reports).*
Note: The table represents the key market opportunities for pseudocereal-based products, driven by major consumer trends toward health, wellness, and sustainability. The high projected growth rates indicate significant commercial potential.

Table 4.
Market outlook for quinoa- and amaranth–based product categories (2023–2030).

consumer base has created expanding niche markets for pseudocereals. For instance, the global quinoa market was valued at USD 90 million in 2022 and is projected to grow at 8%–10% annually [26]. Similarly, the rise of gluten-free diets has driven increased demand for amaranth and buckwheat (see **Table 4**).

Despite these opportunities, policy support – including subsidies, public procurement, and inclusion in food safety nets – remains heavily biased toward rice, wheat, and maize. Incorporating pseudocereals into food security programs and value chains can enhance dietary diversity and improve farmers' incomes.

6.1 Emerging market opportunities for pseudocereals

The growing interest in quinoa and amaranth as functional and nutritionally superior grains has significantly influenced the development of value-added food products across multiple categories. These ancient grains are naturally gluten-free, rich in high-quality proteins, dietary fiber, minerals, and bioactive compounds, making them highly suitable for modern health-oriented diets (see **Table 3**).

The gluten-free bakery segment is driven by the rising prevalence of celiac disease and non-celiac gluten sensitivity, alongside increasing consumer demand for healthier alternatives to refined wheat-based products. Similarly, the plant-based protein category is expanding rapidly due to sustainability concerns and the global shift toward plant-forward diets, positioning quinoa and amaranth as attractive protein sources with balanced amino acid profiles.

The ancient grains and superfoods segment caters to wellness-conscious and premium consumers seeking clean-label and minimally processed foods. In the infant nutrition sector, quinoa-based formulations are gaining attention due to their digestibility and hypoallergenic properties, addressing parental concerns regarding early-life nutrition and food safety. Furthermore, the non-dairy beverage market is witnessing the fastest growth, fueled by widespread lactose intolerance and the increasing adoption of vegan lifestyles.

Overall, the projected compound annual growth rates (CAGR) across these product categories highlight the "strong commercial potential" of quinoa- and amaranth-based foods. Their versatility, nutritional advantages, and alignment with contemporary dietary trends make them promising ingredients for future food product innovation and market expansion.

7. Technological and biotechnological innovations

Recent advances in crop breeding, genomics, and biotechnology are reshaping the future of both cereals and pseudocereals. High-yielding and stress-tolerant varieties of maize and rice are under development, while biofortification initiatives – such as zinc-enriched wheat and provitamin A-rich maize – are addressing critical nutritional deficiencies. Concurrently, research on pseudocereals like quinoa and amaranth focuses on expanding their adaptability and enhancing yields beyond their traditional cultivation regions.

In addition, digital agriculture, precision farming, and integrated crop management practices are improving efficiency, resource use, and sustainability across cereal and pseudocereal systems. Cutting-edge genome editing technologies, such as CRISPR, are being explored to further enhance nutritional quality and stress tolerance, offering new pathways to resilient and nutrient-rich crop production.

Scientific advancement plays a crucial role in unlocking the full potential of pseudocereals for modern agriculture and nutrition. Precision breeding and genomics have accelerated crop improvement, with genome sequencing efforts for quinoa (30), amaranth, and buckwheat enabling marker-assisted selection and advanced gene-editing tools such as CRISPR-Cas9. These innovations allow breeders to develop high-yielding, stress-tolerant varieties with fewer anti-nutritional factors far more rapidly than through conventional methods. Complementing this, climate-smart agronomic research – ranging from optimizing planting densities and refining water-management practices to designing efficient intercropping systems like maize–amaranth combinations – helps improve resource use efficiency and boost overall system productivity. Equally important is progress in post-harvest technology, particularly the development of effective methods to remove saponins from quinoa and the creation of stable, value-added pseudocereal ingredients for the food industry. Together, these scientific advancements strengthen the economic viability, marketability, and sustainability of pseudocereal cultivation on a global scale.

8. Policy, sustainability, and food security perspectives

Ensuring food and nutritional security in the twenty-first century requires an integrated policy approach that gives balanced importance to both staple cereals and underutilized pseudocereals. Diversifying national and household food baskets by incorporating pseudocereals into public distribution systems, mid-day meal schemes, and other nutrition programs can significantly improve dietary quality and resilience. At the same time, sustainability goals can be advanced by promoting water-efficient and climate-resilient crops such as millets and quinoa, especially in drought-prone and resource-scarce regions. Supportive market and trade policies are essential to incentivize smallholder farmers to adopt these crops by ensuring fair prices, procurement mechanisms, and value-chain development. Furthermore, greater research investment is needed to enhance the agronomic performance, processing technologies, and global supply chains of pseudocereals so that they can contribute meaningfully to long-term food and nutritional security.

9. Conclusion

In conclusion, reimagining the role of grains in the twenty-first century is an imperative, not a choice. The evidence presented in this chapter underscores that an over-reliance on a few major cereals is a strategy fraught with risk in the face of climate change, resource depletion, and burgeoning nutritional crises. The future of food security and nutritional well-being hinges on our ability to innovate and diversify. Pseudocereals, once relegated to the margins of agricultural research and policy, are now poised to play a central role in this new vision. They offer a blueprint for designing food systems that are not only productive but also nourishing and resilient.

Realizing this potential, however, requires a concerted, multi-pronged effort: accelerating genomic research to develop high-yielding, climate-smart varieties of both cereals and pseudocereals; creating economic incentives for farmers to adopt diversified cropping systems; and fostering consumer-driven markets through awareness and innovative food products. The journey ahead is one of collaboration across disciplines and sectors. By championing innovation and embracing the full spectrum of the world's grain diversity, we can transform our agricultural landscapes and food plates, ensuring that the foundation of human sustenance is secure for generations to come.

Acknowledgments

The author gratefully acknowledges the use of AI-based language assistance tools, specifically Deep Seek and ChatGPT (OpenAI), for language polishing and improving the readability of this chapter. All intellectual content, interpretations, analyses, and conclusions presented in the chapter are entirely human-generated, and the AI tool was used only for minor editorial refinement. The author also confirms that all references and scientific information have been manually verified to ensure accuracy and integrity.

Author details

Pramod Kumar Sharma[1*], Kadadanamari S. Shashidhar[1], S. Bhuvaneshwari[2], Mahipal Choubey[1] and Ajeet Singh[1]

1 Banaras Hindu University, Varanasi, India

2 ICAR-Indian Institute of Vegetable Research, Varanasi, India

*Address all correspondence to: shashi81@bhu.ac.in

References

[1] Guns DJ. Germs, and Steel: The Fates of Human Societies. New York: W. W. Norton & Company; 1997

[2] Shewry PR, Hey SJ. Do we need to worry about eating wheat? Nutrition Bulletin. 2015;**40**(1):6–13

[3] FAO. The State of Food Security and Nutrition in the World 2022. Rome: Food and Agriculture Organization of the United Nations; 2022

[4] Pingali P. Green Revolution: Impacts, limits, and the path ahead. Proceedings of the National Academy of Sciences. 2012;**109**(31):12302–12308. DOI: 10.1073/pnas.0912953109

[5] Tilman D, Balzer C, Hill J, Befort BL. Global food demand and the sustainable intensification of agriculture. Proceedings of the National Academy of Sciences. 2011;**108**(50):20260–20264. DOI: 10.1073/pnas.111643710

[6] Lobell DB, Cassman KG, Field CB. Crop yield gaps: Their importance, magnitudes, and causes. Annual Review of Environment and Resources. 2009;**34**:179–204. DOI: 10.1146/annurev.environ.041008.093740

[7] Bouis HE, Saltzman A. Improving nutrition through biofortification: A review of evidence from HarvestPlus, 2003 through 2016. Global Food Security. 2017;**12**:49–58. DOI: 10.1016/j.gfs.2017.01.009

[8] World Health Organization (WHO). The State of Food Security and Nutrition in the World 2021: Transforming Food Systems for Food Security, Improved Nutrition and Affordable Healthy Diets for All. Rome: FAO, IFAD, UNICEF, WFP and WHO; 2021

[9] Bhargava A, Srivastava S. Quinoa: Botany, Production and Uses. Wallingford: CABI; 2013

[10] Vega-Gálvez A, Miranda M, Vergara J, Uribe E, Puente L, Martínez EA. Nutrition facts and functional potential of quinoa (Chenopodium quinoa willd.), an ancient Andean grain: A review. Journal of the Science of Food and Agriculture. 2010;**90**(15):2541–2547. DOI: 10.1002/jsfa.4158

[11] Jacobsen SE. The Worldwide Potential for Quinoa (Chenopodium quinoaWilld.). Food Reviews International. 2003;**19**(1–2):167–177. DOI: 10.1081/FRI-120018883

[12] Repo-Carrasco-Valencia R, Serna LA. Quinoa (Chenopodium quinoa, Willd.) as a source of dietary fiber and other functional components. Ciência E Tecnologia De Alimentos. 2011;**31**(1):225–230. DOI: 10.1590/s0101-20612011000100035

[13] Ruales J, Nair BM. Nutritional quality of the protein in quinoa (Chenopodium quinoa, Willd) seeds. Plant Foods for Human Nutrition. 1992;**42**(1):1–11. DOI: 10.1007/BF02196067

[14] FAO. The State of Food and Agriculture 2020. Rome: Food and Agriculture Organization of the United Nations; 2020

[15] Foley JA, Ramankutty N, Brauman KA, Cassidy ES, Gerber JS, Johnston M, Mueller ND, O'Connell C,

Ray DK, West PC, Balzer C, Bennett EM, Carpenter SR, Hill J, Monfreda C, Polasky S, Rockström J, Sheehan J, Siebert S, Tilman D, Zaks DPM. Solutions for a cultivated planet. Nature. 2011;**478**(7369):337–342. DOI: 10.1038/nature10452

[16] IPCC. Climate Change 2022: Impacts, Adaptation and Vulnerability. Cambridge: Cambridge University Press; 2022

[17] Khoury CK, Bjorkman AD, Dempewolf H, Ramirez-Villegas J, Guarino L, Jarvis A, Rieseberg LH, Struik PC. Increasing homogeneity in global food supplies and the implications for food security. Proceedings of the National Academy of Sciences of the United States of America. 2014;**111**(11): 4001–4006. DOI: 10.1073/pnas.1313490111

[18] FAOSTAT: Food and Agriculture Data [Internet]. Rome: Food and Agriculture Organization of the United Nations; 2022. Available from: https://www.fao.org/faostat/en/ [Accessed: 2025-September-10]

[19] FAOSTAT: Food and Agriculture Data [Internet]. Rome: Food and Agriculture Organization of the United Nations; 2023. Available from: https://www.fao.org/faostat/en/ [Accessed: 2025-September-10]

[20] Nascimento AC, Mota C, Coelho I, Gueifão S, Santos M, Matos AS, Gimenez A, Lobo M, Samman N, Castanheira I. Characterisation of nutrient profile of quinoa (Chenopodium quinoa), amaranth (Amaranthus caudatus), and purple corn (Zea mays L.) consumed in the North of Argentina: Proximates, minerals and trace elements. Food Chemistry. 2014;**148**:420–426. DOI: 10.1016/j.foodchem.2013.09.155

[21] Alexandratos N, Bruinsma J. World Agriculture Towards 2030/2050: The 2012 Revision. Rome: Food and Agriculture Organization of the United Nations; 2012

[22] Alvarez-Jubete L, Arendt EK, Gallagher E. Nutritive value of pseudocereals and their increasing use as functional gluten-free ingredients. Trends in Food Science & Technology. 2009;**21**(2):106–113. DOI: 10.1016/j.tifs.2009.10.014

[23] Fghire R, Anaya F, Ali OI, Benlhabib O, Ragab R, Wahbi S. Physiological and photosynthetic response of quinoa to drought stress. Chilean Journal of Agricultural Research. 2015;**75**(2): DOI: 10.4067/S0718–58392015000200006

[24] Zhaoa C, Liu B, Piao S, Wang X, Lobell DB, Huang Y, Huang M, Yao Y, Bassu S, Ciais P, Durand J-L, Elliott J, Ewert F, Janssens IA, Tao L, Lin E, Liu Q, Martre P, Müller C, Peng S, Peñuelas J, Ruane AC, Wallach D, Wang T, Donghai W, Liu Z, Zhu Y, Zhu Z, Asseng S. Temperature increase reduces global yields of major crops in four independent estimates. Proceedings of the National Academy of Sciences. 2017;**114**(35):9326–9331. DOI: 10.1073/pnas.1701762114

[25] Jarvis DE, Yung Shwen H, Lightfoot DJ, Schmöckel SM, Li B, Borm TJA, Ohyanagi H, Mineta K, Michell CT, Saber N, Kharbatia NM, Rupper RR, Sharp AR, Dally N, Boughton BA, Woo YH, Gao G, Schijlen EGWM, Guo X, Momin AA, Negrão S, Al-Babili S, Gehring C, Roessner U, Tester M. The genome of Chenopodium quinoa. Nature. 2017;**542**

(7641):307–312. DOI: 10.1038/nature21370

[26] Alvarez-Jubete L, Arendt EK, Gallagher E. Nutritive value and chemical composition of pseudocereals as gluten-free ingredients. International Journal of Food Sciences and Nutrition. 2009;**60**(s4):240–257

Section 3

Sustainable Cultivation and Nutritional Innovation in Underutilized Crops

Chapter 4

Sustainable Cultivation Practices, Nutritional Optimization, and Food Processing Innovations for Bambara Groundnut in the Sub-Saharan and Namibian Context

Lydia Ndinelao Horn, Felicitas Mantjodi Fwanyanga, Justina Itaveleni Shimwafeni, Beatrix Namulo Alweendo, Maria Kapembe and Kuume Benjamen Pendapala Enguwa

Abstract

Bambara groundnut (Vigna subterranea (L.) Verdc), an indigenous legume crop in sub-Saharan Africa, is a crucial source of protein. It is known for its ability to form symbiotic relationships with Rhizobia strains that fixes atmospheric nitrogen through the Biological Nitrogen Fixation (BNF) process. In Namibia, it ranks as the second most important legume after cowpea and is grown in the northern communal regions of the country where it is also consumed. The crop's nutritional value beats that of many staple foods, providing comprehensive protein in addition to a significant amount of crucial nutrients, such as magnesium and iron, which help fight the major crisis of malnutrition in Africa. Despite its nutritional value, it has been neglected in scientific research and agricultural development programs. The crop faces challenges such as biotic and abiotic stresses, socio-economic constraints, and low yields. Its integration into sustainable food systems is urgently needed, given the challenges of climate change, food insecurity, and nutritional deficiencies. Improving Bambara groundnut requires identifying and selecting genotypes with desirable traits such as early maturity, drought tolerance, larger seed size, and higher yield. Significant genotypic and phenotypic variability exists within landrace populations. This chapter aims to examine sustainable cultivation practices, nutritional optimization, and food processing innovations in Namibia and elsewhere in the world. It emphasizes the need for improved agronomic practices, conservation agriculture, and breeding activities. The review also examines methods to broaden genetic diversity, utilize genomic resources, and employ high-throughput phenotyping to enhance the utilization of genetic resources in pre-breeding programs.

IntechOpen

Keywords: Bambara groundnut, Sub-Saharan Africa, namibia, food security, crop improvement

1. Introduction

Bambara groundnut (*Vigna subterranea* [L.] Verdc.) is a nitrogen-fixing legume known for its ability to form symbiotic relationships with Rhizobia strains [1]. Although it is underutilized and largely overlooked both in Africa and globally, it remains an essential crop for many local farmers. Native to Sub-Saharan Africa, it is widely consumed across the region. While subsistence farmers cultivate Bambara groundnut in considerable quantities, information such as their varietal preferences, seed sources, production practices, and challenges remains unknown to crop breeders [2].

In Namibia, Bambara groundnut ranks as the second most important legume after cowpea and is grown in the northern communal regions of the country [3]. Unlike cowpea, this crop is, in most cases, grown as a monocrop in small plots in selected, suitable areas and grows well at an optimum pH of 5.0–6.5 [4]. It is also one of the few leguminous crops cultivated for human consumption that is well adapted to the country's increasingly dry climatic conditions [5, 6].

In the face of increasing climate challenges and acute malnutrition across Sub-Saharan Africa, Bambara groundnut emerges as an invaluable yet neglected crop [7, 8]. The Bambara groundnut, a resilient African indigenous legume crop, possesses both valuable agronomic and nutritional qualities that qualify it as an outstanding solution for addressing food insecurity and promoting sustainable agriculture [9]. Despite this, Bambara groundnut remains underutilized in formal agricultural systems across Sub-Saharan Africa, including Namibia. Its integration into sustainable food systems is urgently needed, given the challenges of climate change, food insecurity, and nutritional deficiencies.

2. Importance of Bambara groundnut that is still not exploited in Namibia and Africa in general

The Bambara groundnut is a vital crop in African agriculture. Although the crop can perform under challenging conditions while offering remarkable nutritional benefits, it remains overlooked in research and political discussions [7, 10]. In terms of climatic resilience, the crop thrives under devastating drought conditions, yielding 0.85–3 t/ha and outperforming other crops such as maize and beans [9]. Similarly, the Bambara groundnut's nutritional value surpasses that of many staple foods, providing complete protein in addition to a considerable amount of essential nutrients, such as magnesium and iron [9], which help combat the predominant crisis of malnutrition in Africa. Moreover, this presents an economic opportunity for food processing industries through value addition and product development, improving the livelihoods of smallholder farmers. Despite these benefits, there is minimal attention directed toward funding scientific research on Bambara groundnut in Sub-Saharan Africa, underscoring the neglect of such a valuable local crop [10].

3. Cultivation and agronomic traits of importance

3.1 Crop morphology and phenology

Bambara groundnut is a self-pollinating legume that is morphologically similar to the peanut (*Arachis hypogaea* L.) in terms of underground pod development. It is a low-growing annual plant with creeping stems, often forming a crown at the soil surface. The plant displays considerable morphological variation, reflecting its diverse genetic pool across geographic regions [11].

The leaves are trifoliate, alternate, and glabrous, with elliptical to oblanceolate leaflets ranging from 3 cm to 10 cm in length and 1 to 5 cm in width. The petiole can extend up to 30 cm and is thickened at the base. Flowers are typically papilionaceous, bisexual, and yellow or whitish. They are borne on long, hairy peduncles that emerge from the stem nodes. Following fertilization, the flower stalk (peduncle) elongates and directs the developing fruit toward the soil, a process known as geocarpy, similar to that of the groundnut. Pods develop underground and are usually indehiscent, containing one or two seeds [4, 12].

The seeds themselves are round to oval in shape, with various coat colors, including cream, red, black, and speckled patterns. Farmers often use seed color to differentiate local landraces, which may be associated with consumer preferences and marketability [11].

Phenologically, Bambara groundnut exhibits both determinate and indeterminate growth habits, depending on the genotype. This means flowering and pod development may occur continuously over an extended period, especially in indeterminate types. Flowering typically begins 30–55 days after planting, while pod development and maturity occur over the next 60–90 days [4, 12].

3.2 Environmental adaptability and resilience

Climate change poses a serious threat to crop production, especially in Africa, with its extensive arid and semi-arid regions. The cultivation of Bambara groundnut serves as a strategy to mitigate the effects of climate change [9]. Unlike other heavy-feeding crops, this crop has evolved to thrive in a broad range of agro-ecological zones, including semi-arid regions with low and erratic rainfall, poor soil fertility, and high temperature variability [12]. Its resilience is attributed to both genetic and physiological adaptations, making it a suitable crop for sustainable agriculture in marginal environments.

One of its most important agronomic advantages is drought tolerance. Bambara groundnut requires only 300–700 mm of rainfall during the growing season, significantly less than other legumes like soybeans or common beans [11]. Optimal growth occurs at 19–30°C, but the crop can withstand high daytime temperatures and cold nighttime temperatures, especially in arid zones such as Botswana and Namibia. However, growth is severely inhibited at temperatures above 35°C or below 15°C. Freezing temperatures at any stage of growth are fatal, which limits its cultivation to frost-free environments [13].

Bambara groundnut is also highly soil-adaptable, thriving in well-drained sandy loam soils with a pH between 5.0 and 6.5. It tolerates acidic and nutrient-deficient soils due to its ability to fix nitrogen symbiotically. Studies show that it can fix between 32 and 81 kg of atmospheric nitrogen per hectare, thereby enhancing soil fertility and reducing the need for synthetic fertilizers in intercropping or rotation systems [4].

Genotype-by-environment interactions are pronounced in Bambara groundnut, with different accessions showing localized adaptation. For instance, genotypes from arid sub-Saharan zones exhibit deep taproots and conservative water-use strategies, while those from wetter regions display shallower, highly branched root systems. Such root system architecture is critical for water capture and drought avoidance.

3.3 Key agronomic traits for improvement

One of the most outstanding agronomic traits of Bambara groundnut is its ability to thrive in nutrient-poor soils with a pH of 5.0–6.5 and survive droughts, even in areas that receive an average of 500–600 mm of rainfall per year [7, 10]. Furthermore, the crop can tolerate drought and survive in areas where other leguminous crops, such as groundnut and cowpea, fail. It has been reported [4] that the drought tolerance of Bambara groundnut is due to the plant's ability to maintain leaf turgor pressure through a combination of osmotic adjustment, reduction in leaf area, and effective stomatal control.

In addition, as a legume, Bambara groundnut plays a crucial role in crop production and sustainable agriculture in general, as it forms hosts for soil bacteria that facilitate the nitrogen fixation process and thus improve soil fertility [2, 10]. This helps to reduce dependence on the use of inorganic fertilizers, a cost-saving strategy that is particularly important for smallholder farmers with limited access to expensive inputs. Furthermore, Bambara groundnut plays a critical role in crop diversification, for example, intercropping with cereals, which leads to a yield increase of 20–40% compared to mono-cropping [10, 14].

As yield variability remains a challenge for most farmers due to their reliance on uncertified planting materials with inconsistent performance [2], there is a need to use improved seeds and adopt more context-specific, best management practices.

Improving Bambara groundnut requires identifying and selecting genotypes with desirable traits such as early maturity, drought tolerance, larger seed size, and higher yield. Significant genotypic and phenotypic variability exists within landrace populations, and these variations have been confirmed through morphological, agronomic, and genomic assessments [15]. According to [16], high heritability has been recorded for traits such as pod number, seed weight, and flowering time, indicating potential for selection-based improvement.

Furthermore, genomic tools, such as genome-wide association studies (GWAS), have identified several marker-trait associations related to seed coat color, days to flowering, pod width, and terminal leaf traits [17]. Trait preferences identified through participatory rural appraisals, such as early maturity, high yield, and drought tolerance, should inform breeding objectives [2]. Therefore, integrating farmer knowledge, modern molecular tools, and agro-ecological considerations will be key to developing improved Bambara groundnut varieties tailored to different regions and growing conditions.

4. Soil management practices for Bambara groundnut

4.1 Soil fertility requirement

Bambara groundnut can grow in a wide range of agro-ecological zones that differ in soil conditions. The crop can grow well even in soils characterized as poor or infertile [18]. However, its productivity can be greatly improved if cultivated on healthy and fertile soils and supplemented with fertilizer. Several studies have examined the application of synthetic fertilizers and a combination of synthetic and organic (organomineral) fertilizers for Bambara groundnut production. For example, Temegne et al. [19] reported that applying phosphate (KH_2PO_4) at a concentration of 1,000 μM Pi increased Bambara groundnut growth and leaf P content, which was attributed to the N and P interaction in the root zone. Similarly, a study by Ramolemana et al. [20] on "The phosphorus and nitrogen nutrition of Bambara groundnut in Botswana soils", results showed that under soils with low P, the application of P significantly improved Bambara groundnut yield and yield parameter such as number of pods, pod dry weight and dry matter production. However, these positive results were only observed when the P fertilizer was applied within two weeks after sowing. Legumes are known as phosphorus-loving crops [21]. P is needed in significant quantities and is involved in the key functions of several plants, including energy transfer, photosynthesis, sugar and starch transformation, nutrient movement in the plant, and transfer of genetic characteristics [22].

Like many other legumes, Bambara groundnut fixes atmospheric nitrogen through the biological nitrogen fixation (BNF) process. Legumes, including Bambara groundnut, can fix up to 11–20 kg of N/ha, but this is not achievable in the tropics because of low soil fertility and poor farming practices [23–25]. On the other hand, it has been reported that a high nitrogen level in the soil can lead to excessive vegetative growth instead of pod formation, resulting in a subsequent low yield. When nitrogen is added late during the growing season, it will also result in poor nodulation [4]. In fact, [26] found that applying 2 t/ha of ground cow manure produced significantly heavier dry nodules than 3, 4, and 5 t/ha. Therefore, nitrogen application for Bambara groundnut should be done with extra care.

The BNF process is also very important for improving soil fertility and supplying other non-leguminous crops with the nitrogen left in the soil after the legume is harvested. This BNF process in legumes is carried out by forming a symbiotic association with a group of soil bacteria, commonly known as rhizobia, which naturally fix atmospheric nitrogen by forming specialized structures on the roots, known as nodules.

4.2 Soil structure and drainage

The choice of land and soil type for Bambara groundnut cultivation is crucial, as this can profoundly affect the crop's growth and yield. The ideal site for Bambara groundnut should be fairly level to avoid water run-off. In addition, a well-drained site should be selected to prevent pod rotting caused by waterlogging. Moreover, stony areas should be avoided to ensure good pod development. Therefore, sandy loam soils are ideal for cultivating Bambara groundnut, as they ensure healthy pod development and easy harvesting [4].

4.3 Water management

Bambara groundnut can produce a significant yield under moderate or extreme drought stress. However, research has shown that drought stress on the crop leads to reductions in pod dry matter and, ultimately, reduced grain yields, particularly if it occurs during the flowering stage. The total minimum rainfall requirement for Bambara groundnut is estimated to be 300 mm [26], but it performs better under annual rainfall levels between 750 mm and 1,400 mm [27]. Excessive rainfall conditions, especially during harvest, result in drastic yield losses [10]. During drought seasons, several management practices, such as the integration of landscape and field-scale conservation technologies, have been reported to positively affect dryland farming, including Bambara groundnut production in Sub-Saharan Africa, particularly in arid and semi-arid regions with limited water resources [28].

5. Available varieties and their sustainability

The Bambara groundnut crop is genetically diverse and offers opportunities for sustainable agricultural practices. In Sub-Saharan Africa, farmers typically rely on different landraces, which usually differ in seed color, size, and maturity [2]. This variability is particularly pronounced in Namibia. Although locally adapted varieties produce moderate yields of 0.85 t/ha, with a yield potential of 3 t/ha [9, 10], there is a need for breeding initiatives to develop improved varieties, such as early-maturing and high-yielding cultivars [2]. The cultivation of Bambara groundnut is ecologically, economically, and socially sustainable. It has higher water use efficiency and lower input costs compared to other crops and is mostly cultivated by women [10]. However, the loss of various genetic traits poses a major threat to valuable Bambara landraces, hence the need to conserve them while bridging the gap between seed banks and breeding programs [2].

6. Bambara groundnut: Production opportunities and challenges for local communities

6.1 Production challenges

Although most rural communities in both Namibia and Sub-Saharan Africa as a whole benefit significantly from the Bambara groundnut, challenges are inevitable. Several biotic and abiotic constraints hindering Bambara groundnut production have been documented. Biotic stressors include fungal, bacterial, and viral diseases, insect pests such as bruchids (*Callosobruchus maculatus*), and nematode infestations [2, 11]. These reduce both yield and quality. Although some diseases can be managed through crop protection chemicals and practices such as crop rotation, burning of infected residues, or the use of organic remedies, smallholder farmers often lack access to such interventions [28].

Abiotic stresses, such as erratic rainfall, extreme temperatures, and low soil fertility, are prevalent in Bambara-growing regions, including the sub-Saharan region [3]. Climate change exacerbates these stresses by altering rainfall patterns, reducing planting windows, and increasing the frequency of drought episodes [11]. Additionally, most Bambara landraces are photoperiod-sensitive, requiring short day lengths for optimal pod filling. Under long photoperiods, pod development is significantly impaired [28]. The simultaneous occurrence of heat and drought stress further complicates crop performance and may be governed by overlapping genetic mechanisms [17].

Equally important, socio-economic challenges have also been reported. Despite its resilience and nutritional potential, Bambara groundnut remains underutilized due to socio-economic barriers. It is often perceived as a "poor man's crop" and is cultivated mainly for household consumption rather than commercial purposes [11, 12]. This perception has led to minimal investment in research, extension services, and input systems. Furthermore, there is a lack of well-structured value chains, formal seed systems, and marketing infrastructure. Farmers rely heavily on saved seeds and informal markets, contributing to inconsistent quality and low productivity [3].

Additionally, traditional beliefs have stigmatized the crop as a woman's crop or even as one that brings bad luck, thereby limiting male participation and broader community support [13]. While these views are gradually shifting, the absence of national breeding programs, seed certification systems, and policy inclusion continues to hinder its expansion and development.

In Namibia, Bambara groundnut is cultivated by smallholder farmers primarily for household consumption. The crop is well-adapted to the country's semi-arid environment, but the lack of improved varieties, access to inputs, and consistent extension services has constrained its productivity [10]. Studies have shown that yields in most farming systems are less than 1 ton per hectare, and the crop is typically grown on plots smaller than one hectare using traditional practices [2, 5]. Strategies to overcome such challenges include the formation of farmers' associations to promote market accessibility, the introduction of economically viable processing machinery such as dehullers, and the formulation and implementation of relevant policies.

6.2 Production opportunities

Notwithstanding these constraints, Bambara groundnut offers considerable opportunities for sustainable agriculture, especially in marginal and drought-prone areas. Its ability to grow under low-input conditions, fix atmospheric nitrogen, and produce stable yields under adverse conditions makes it an ideal crop for climate-smart agriculture [11, 28]. It is highly nutritious, containing balanced levels of protein, carbohydrates, and micronutrients, and is sometimes described as a "complete food" [13].

Bambara groundnut's adaptation to semi-arid climates, combined with its potential role in crop diversification, positions it as a strategic crop for enhancing food and nutritional security in Sub-Saharan Africa. With rising global interest in plant-based proteins and sustainable crops, Bambara groundnuts could be promoted as a health food in niche global markets [11]. Genetic diversity within landraces provides ample breeding material to improve drought tolerance, early maturity, and yield potential [12, 17].

Studies have shown that the inclusion of Bambara flour in school feeding programs has improved the nutrition of children in Malawi. Meanwhile, Namibian farmers, particularly those from the Omusati region, have achieved ideal yields with Bambara, even when the overall harvest for other crops was poor [2]. This also applies to the economic benefits and opportunities, as processed Bambara products yield an excellent profit compared to raw, unprocessed beans [10].

7. Nutritional value and opportunity for product formulation

The high nutritional composition of Bambara groundnut makes it the best choice for product development and value-added processing industries. Bambara is high in protein (23.6%), carbohydrates (64.4%), fat (6.5%), fiber (5.5%), and important minerals such as calcium, iron, and magnesium [9, 29]. According to [30], Bambara groundnut has a higher protein quality, recorded at 80%, which is more than that of groundnut (65%) and cowpea (64%). Bambara groundnut can be mixed into cereal products as an additive. It also has a low glycemic content, which makes it more suitable for products intended for diabetics [2]. Studies have shown that adding 15% of Bambara flour to bread can increase the protein content by 40% without compromising the bread's quality [29]. However, the processing of Bambara groundnut is quite laborious, for example, soaking and grinding/milling of the seeds, which highlights the need for advanced technologies to streamline the process and ultimately lead to lower production costs [10].

7.1 Nutritional profile and significance

Bambara groundnut is considered a complete food. The legume's excellent nutritional profile is characterized by the presence of various macro and micronutrients, which vary based on varieties and geographical conditions. The carbohydrate content makes Bambara groundnut an energy-dense food compared to other legumes, such as cowpea [31]. The rich carbohydrate content ranges between 49% and 63% of its dry weight [9, 32]. Due to its high content of complex starches and dietary fiber, Bambara groundnuts have a low glycemic index and provide long-lasting energy. This can help control blood sugar levels and reduce the risk of diabetes and obesity, which are major health issues in Namibia.

The protein content ranges between 9% and 40%, depending on the variety and other geographical conditions [9, 33]. The amino acid profile of Bambara groundnut shows high levels of glutamic acid, leucine, and lysine, while methionine is the limiting amino acid. This makes Bambara groundnut a good complementary protein source to cereals, which are often deficient in lysine but rich in methionine [9]. The protein profile in Bambara groundnut can serve as an affordable protein substitute that complements instant porridges for school-going children and snacks for pregnant women.

Although Bambara groundnut is praised for its excellent nutritional profile, the same does not hold true for its fat content. The fat content in Bambara groundnut can range from as low as 1.4% to as high as 12% [9, 34]. The fat content in Bambara groundnut has been reported to be lower compared to other legumes, such as soybeans and peanuts, which have fat contents ranging between 20% and 50%, respectively [6].

The fatty acid composition in Bambara groundnut is mainly made up of oleic acid and linoleic acid (omega-6) [9]. Oleic acid and linoleic acid in Bambara groundnut have the potential to reduce heart disease, support brain development in children, combat chronic illnesses caused by malnutrition, and limit diabetes-related complications [35], which are raising concerns in Namibia.

Bambara groundnut is rich in various micro-nutrients, which can address the issue of food insecurity that has proven to be a challenge for communities affected by malnutrition in Namibia. The legume is particularly rich in calcium (95.8–360 mg per

100 g), iron (4.9 mg–48 mg per 100 g), potassium (11.44–19.35 mg per 100 g), and sodium (2.9–12.0 mg per 100 g) [34]. Other studies have reported traces of zinc, magnesium, and phosphorus in Bambara groundnut. The combined presence of these minerals contributes to a range of functions in body development, including bone health, cardiovascular health, and boosting immune function.

7.2 Challenges to nutritional utilization

Despite the bountiful nutritional profile of the Bambara groundnut, several factors limit the bioavailability of the legume's nutrients.

7.2.1 Antinutritional factors (ANFs)

Despite its nutritional benefits, the accessibility of the dietary value of Bambara groundnut is limited by the presence of ANFs, which interfere with the digestion and bioavailability of essential nutrients [36]. ANFs are biological compounds naturally present in legumes that bind to essential nutrients, thus limiting the body's uptake and digestibility. Common ANFs found in Bambara groundnuts include tannins, trypsin inhibitors, and phytic acid [31].

Tannins are polyphenolic compounds that form complexes with Bambara groundnut, thus limiting protein availability [37]. There is a common notion [38] that the relationship between tannins and seed coat color plays a role in the taste of Bambara groundnuts. Tannins, which are more prevalent in darker-colored Bambara groundnuts, contribute to the bitter and astringent taste of the legume [9].

Trypsin inhibitors are also a group of ANFs that have been reported to significantly limit protein digestibility in raw or unprocessed Bambara groundnuts. These inhibitors limit the activity of trypsin enzymes, thus negatively impacting the digestion and absorption of proteins in the human body [39].

Phytic acid is another common ANF present in high volumes in Bambara groundnuts. Its presence is linked to the reduction in calcium accessibility [38], the inhibition of pepsin, lipase, and amylase activity [39], reduced iron bioavailability, and the formation of crosslinks with dietary proteins and starch [9].

7.2.2 Hard-to-cook (HTC) phenomenon

The HTC phenomenon is a characteristic that causes the hardening of the seed coat in Bambara groundnut when the legume is stored at excessively high temperatures (>25°C) and/or high levels of humidity (>65%) [9]. The hard seed coat of Bambara groundnut contributes to an increase in cooking time (up to four hours), which subsequently increases energy consumption and reduces milling efficiency [40]. This HTC phenomenon discourages consumption of the legumes, and cooking proves to be a challenge for rural households in Namibia that heavily depend on traditional cooking methods.

Nonetheless, the HTC can be reduced through processing methods, mainly fermentation and soaking. Significant research indicates that the HTC component in Bambara groundnut can be reduced by boiling the legumes in alkaline salts. Boiling Bambara groundnuts in a 0.5 g/100 mL solution of sodium bicarbonate, sodium chloride, and magnesium sulfate for 30 minutes had various nutritional effects [31].

Bambara groundnuts boiled in sodium bicarbonate had higher carbohydrate and crude fiber content, which suggests the effectiveness of the alkaline salt in softening the hardened seed coat. Bambara groundnuts boiled in magnesium sulfate had higher protein, potassium, and iron content, which suggests the effective destruction of ANFs. Bambara groundnuts boiled in sodium chloride had the highest fat and phosphorus content.

7.2.3 Processing techniques

In Namibia, Bambara groundnut is traditionally processed and consumed through various methods such as boiling, sun drying after harvesting, freshly cooking, or combining with other grains and cooking as food [41]. Sun drying of the pods is mainly practiced, and because the drying process lacks control, sometimes over-drying of pods occurs, leading to poor shelling output [42]. According to Gerrano et al. [4], processing methods not only affect the nutritional value of Bambara groundnuts but also have the potential to enhance their optimal utilization as a legume. There is a variety of processing methods for this legume, such as fermentation, germination, and soaking; however, there are no studies in Namibia that focus on the effects these methods have on Bambara groundnuts. In other studies, however, it has been proven that these methods can effectively break down non-digestible oligosaccharides, which are associated with the production of flatus, into simpler carbohydrates and enhance protein digestibility [36]. Soaking of Bambara groundnuts is done either in hot or cold water at varying time intervals. Studies have recognized a greater loss of trypsin inhibitors and tannins during immersion in hot water, whereas a contradictory observation was made for phytate [36]. A study by Adeleke et al. [37] assessed that soaking is more effective in reducing the ANFs of the samples, with the control seeds exhibiting lower values than the newly developed seeds. On the other hand, boiling time may be required to make the seeds soft enough to eat, which may require high energy during cooking, unlike other legume crops such as cowpea [4].

There is little information available on how long it takes for Bambara grains to cook fully; therefore, there is a need for research to focus on cooking time and the reduction of overall functional utility of Bambara. This is because the hardening of seeds causes a decrease in nutritional quality, and extended boiling leads to reduced nutritional quality and digestibility. The hardening of legume seeds is observed to cause a decrease in nutritional quality, while extended boiling leads to reduced protein digestibility [4].

7.3 Strategies for nutritional optimization

In view of standard processing methods, recent studies have discovered that inoculating Bambara groundnut with bacterial strains is likely to improve nutritional content. Research [31] found that Namibian-grown Bambara groundnut varieties inoculated with *Bradyrhizobium* bacteria displayed unique nutritional traits. In the study, red-colored Bambara groundnuts inoculated with *Bradyrhizobium* had higher fat, magnesium, copper, and iron content than red Bambara groundnuts not inoculated with *Bradyrhizobium* bacteria. A specific bacterial strain, such as *Bradyrhizobium*, can increase the bioavailability of minerals in legumes by decreasing the amount of ANFs, such as phytic acid, which bind to minerals and reduce their absorption [43, 44]. Brown-colored Bambara groundnuts inoculated with *Bradyrhizobium* had protein content twice as high as brown Bambara groundnuts not inoculated. Bradyrhizobia form a symbiotic relationship with the root nodules of legumes, which increases the nitrogen content and has a significant impact on the protein content.

7.4 Integration into local diets: An example for Namibia

There have been various value-added products developed from Bambara groundnuts around the globe. In Zambia, Bambara groundnut flour is used in canned food; in Benin, dumplings are made; and in East Africa, coffee, milk, noodles, and baby porridge [31] are manufactured. Despite its versatile use, Namibia has only reported the traditional uses of Bambara groundnut to date. Researchers found, through surveys from Kavango East, that Bambara groundnut is used to prepare the following traditional dishes: a boiled snack; "mugombakapindi," which is a stiff porridge made from mahangu flour and Bambara groundnut; "shikuvira," made from roasted Bambara groundnut and peanut flour; and "rutura," which is an instant porridge prepared by mixing warm water with roasted Bambara groundnut flour [41].

Given its high protein content, Bambara groundnuts can be processed into flour and blended with staple foods, such as porridge, to improve the nutritional quality of meals, particularly in school feeding programs, kitchen soups, and orphanage homes in Namibia. The excellent protein content of Bambara groundnuts allows for their versatile use in the manufacturing industry. Proteins in Bambara groundnuts can be isolated, which, in turn, can be used as vegan emulsifiers in sausages and cakes [45]. Furthermore, the starch in Bambara groundnuts can be extracted and utilized as a natural, gluten-, lactose-, and cholesterol-free thickener and stabilizer in various food products, such as sauces, soups, and desserts.

Improving not only the quantity but also the quality of food is essential for sustaining nutritional value [18]. Beyond its role as a food and income source, Bambara groundnut also holds potential for medicinal use and as livestock feed [46].

8. Breeding and genetic selection

In Namibia, efforts have been made to collect and evaluate Bambara groundnut germplasm since independence in the early 2000s, when a project known as BAMFOOD was initiated [47]. However, a formal breeding program for this resilient crop is still lacking [3]. The BAMFOOD (ICA4-CT-2000-30002) project, which took place between 2000 and 2003 through the collaboration of a UK-based organization with SADC member states, collected and assembled several pure lines isolated from the landraces collected from farmers in member states [48]. Recently, preliminary breeding research has been initiated, targeting traits such as drought tolerance, high yield, early maturity, and resistance to pests and diseases [1, 5, 49]. According to Valombola et al. [3], one way of increasing the speed of adoption of new varieties is to give farmers a wide range of novel cultivars to test for themselves in their fields through participatory varietal selection (PVS). PVS is a breeding approach that is used to bring breeders, social scientists, farmers, and extension personnel together in a field to select cultivars based on traits of importance [50]. PVS methods have been well utilized in Namibia [51] and also in other African countries like Ethiopia [50] to test the adoption of several crops such as cowpea and potatoes.

Breeding and genetic selection involve the evaluation of phenotypes and genotypes, as well as the collection of landraces for use in variety improvement [12]. Farmers usually prefer quality traits such as high yield, large leaves, thick and yellow pods, specific seed shapes, bushy growth habits, medium-sized leaves, and nitrogen fixation ability in Bambara groundnut [5, 51]. Many farmers in Namibia indicated that Bambara groundnut plants with larger leaves, bigger pods, and higher yields were mostly preferred [12]. At the same time, in Malawi, small-seeded Bambara groundnut was also preferred and used for relish, while big-seeded Bambara groundnut was used for snacks [8]. Farmers need to have access to high-yielding crops in order to produce enough food for household consumption and also for income generation. Larger pod size is an essential factor for the marketing of Bambara groundnut; this will be required as genetic material in the development of improved varieties, besides an increase in production [12, 16]. It was also found that there is a need to carry out various research studies to determine factors such as the morphological and germination characteristics of Bambara groundnut in order to evaluate the benefits and drawbacks of various genotypes [49]. Several studies on the genetic and agronomic characteristics of the crop displayed the number of flowers, number of leaves, days to maturity, pod development, and yield as traits that vary among the landraces [12, 16]. The environmental conditions were also found to be critical for some growth parameters, but this varies across different locations.

There are no registered or improved Bambara groundnut cultivars known in Namibia, but research dedicated to crop improvement is ongoing, and farmers plant local germplasm, which is mixed and not uniform but exhibits genetic variability [2, 3, 51]. There is, therefore, a need to make Bambara groundnut breeding a priority in order to improve productivity. Selection for increased pod number per plant, pod weight, and seeds per pod can be a strategy for yield improvement and should be considered [2, 5].

9. Improved agronomic practices

The absence of improved agronomic management practices and high-yielding cultivars with pest and disease resistance reduces the production and productivity of Bambara groundnuts for small-scale and commercial farmers in developing countries [42]. In contrast, efforts in Namibia are increasingly focused on promoting improved agronomic practices aimed at enhancing yield, building resilience to climate stress, and improving soil fertility, all within the framework of sustainable agriculture.

Despite these efforts, farmers still rely on indigenous knowledge for crop management and seed preservation [2]. For instance, the use of traditional hoes as a tool for soil preparation has been shown to be effective in increasing growth and yield attributes of Bambara groundnut [52]. Land preparation is especially important for Bambara cultivation because pod formation occurs underground. Well-pulverized soil is believed to aid in the deep penetration of developing pods, thereby supporting optimal growth [4]. Weeding is typically done manually using hand tools such as hoes or, alternatively, by hand-pulling, particularly during the seedling stage [3, 51]. To maximize the benefits of symbiotic nitrogen fixation with rhizobacteria, Bambara groundnut is often intercropped with cereals, either simultaneously or introduced in the middle of the cropping season with these cereals [4]. As in many other African countries, Namibian farmers commonly intercrop Bambara groundnut with staple crops such as maize, pearl millet, and sorghum [2, 5].

10. Conclusion(s)

Bambara groundnut is a drought-tolerant, nutrient-rich legume with great potential to contribute to food security, climate-smart agriculture, and rural livelihoods in Sub-Saharan Africa, especially in Namibia. Despite its adaptability and nutritional benefits, it remains underutilized due to limited research, the unavailability of improved seeds, and socio-economic challenges, among other factors. To tap into Bambara groundnut's full potential, there is a need for targeted breeding programs, improved agronomic practices, value addition, and supportive policies. In addition, integrating traditional knowledge with modern innovations will be crucial to unlocking the role of Bambara groundnut in sustainable agriculture and nutrition.

Acknowledgments

The authors would like to take this opportunity to thank the University of Namibia for granting them the authority to write and submit articles to IntechOpen. The authors further appreciate IntechOpen for accommodating this article.

Conflict of interest

The authors declare no conflicts of interest.

Author details

Lydia Ndinelao Horn[1*], Felicitas Mantjodi Fwanyanga[2], Justina Itaveleni Shimwafeni[3], Beatrix Namulo Alweendo[4], Maria Kapembe[5] and Kuume Benjamen Pendapala Enguwa[6]

1 Zero Emission Research Initiative, Multidisciplinary Research Centre, University of Namibia, Windhoek, Namibia

2 Department of Crop Science, Faculty of Agriculture, Engineering and Natural Resources, University of Namibia, Oshakati, Namibia

3 Mashare Agricultural Development Institute, Ministry of Agriculture, Rundu, Namibia

4 Directorate of Agricultural Research and Training, Ministry of Agriculture, Windhoek, Namibia

5 Department of Food Science and Systems, Faculty of Agriculture, Engineering, and Natural Resources, University of Namibia, Windhoek, Namibia

6 Department of Crop Production and Agricultural Technologies, School of Agriculture and Fisheries Sciences, Oshakati, Namibia

*Address all correspondence to: lhorn@unam.na/lnhorn@yahoo.com

IntechOpen

References

[1] Amwenyo MLD, Horn LN, Hurek T, Reinhold-Hurek B, and Sarkar A. Cultivation constrains of Bambara groundnut in Sub-Saharan Africa: impact of rhizobial inoculants. Plant & Soil. 2025;1–14. DOI: 10.1007/s11104-024-07194-5

[2] Pasipanodya JT, Horn LN, Togarepi C, Sibiya J, Musango R, Achigan-Dako EG. Exploring farmer perspectives on Bambara groundnut [*Vigna subterranea* (L.) Verdc]: Production constraints, preferred varietal traits and implications for varietal development in northern Namibia. Frontiers in Sustainable Food Systems. 2025; **9**:1520677. DOI: 10.3389/fsufs.2025.1520677

[3] Valombola JS, Awala SK, Hove K. Farmers'preferences, seed source, production constraints and improvement need assessment of bambara groundnut (*vigna subterranea* [L.] Verdc.) In northern rural of Namibia. Scientific Papers Series Management, Economic Engineering in Agriculture & Rural Development. 2021;**21**(3):789–798

[4] Gerrano AS, Eifediyi EK, Labuschagne M, Ogedegbe FO, and Hassen AI. Production Practices of Bambara Groundnut. In: Oyeyinka SA, and Ade-Omowaye BIO editors. Food and Potential Industrial Applications of Bambara Groundnut. Cham: Springer; 2021:7–25. DOI: 10.1007/978-3-030-739 20-1_2

[5] Fwanyanga FM, Horn LN, Sibanda T, and Reinhold-Hurek B. Bambara groundnuts [*Vigna subterranea* (L.) Verdc] Production Systems in Kavango East Region, Northern Namibia. Indian Journal of Agricultural Research. 2024;**58**:1087–1093 DOI: 10.18805/IJARe.AF-838

[6] Valombola JS, Labuschagne M, van Biljon A, Horn LN. Determination of optimum dose of lethal dose and growth reduction using gamma-irradiated Bambara groundnut (Vigna Subterranea (L) Verdc.) genotypes. In: Proceedings of the 4th International Electronic Conference on Agronomy; 2–5 December 2024; Basel, Switzerland. Basel: MDPI; 2024. Available from: https://sciforum.net/paper/view/20612

[7] Ayilara MS, A. M. Potentials of underutilized legumes in food security. Frontiers in Soil Science. 2022;**2**:1020193

[8] Mayes S. H. W.Bambara groundnut: An exemplar underutilised legume for resilience under climate change. Planta. 2019;**250**:803–820

[9] Tan XL *et al*, . A.-A. S.-A.Bambara groundnut: An underutilized leguminous crop for global food security and nutrition. Frontiers in Nutrition. 2020;7:601496. DOI: 10.3389/fnut.2020.601496

[10] Majola NG, Gerrano AS, Shimelis H.Bambara groundnut (*Vigna subterranea* [L.] Verdc.) production, utilisation and genetic improvement in Sub-Saharan Africa. Agronomy. 2021; **11**(7):1345

[11] Maphosa Y, Jideani VA, Maphosa L.Bambara groundnut production, grain composition and nutritional value: Opportunities for improvements. The Journal of

Agricultural Science. 2022;**160**(6): 448–458

[12] Muhammad I, Rafii MY, Ramlee SI, Nazli MH, Harun AR, Oladosu Y, Arolu IW.Exploration of bambara groundnut (*Vigna subterranea* (L.) Verdc.), an underutilized crop, to aid global food security: Varietal improvement, genetic diversity and processing. Agronomy. 2020;**10**(6):766

[13] Khan MMH, Rafii MY, Ramlee SI, Jusoh M, Al-Mamun M.Bambara groundnut (*Vigna subterranea* L. Verdc): A crop for the new millennium, its genetic diversity, and improvements to mitigate future food and nutritional challenges. Sustainability. 2021;**13** (10):1–27

[14] Toker P, C. H. The advantages of intercropping to improve productivity in food and forage production–a review. Plant Production Science. 2024;**27** (3):155–169

[15] Olanrewaju OS, Oyatomi O, Babalola OO, Abberton M.Genetic diversity and environmental influence on growth and yield parameters of Bambara groundnut. Frontiers in Plant Science. 2021;**12**(December):1–15

[16] Khaliqi A, Rafii MY, Mazlan N, Jusoh M, Oladosu Y. Genetic analysis and selection criteria in Bambara groundnut accessions-based yield performance. Agronomy. 2021;**11**(8):1634

[17] Uba CU, Oselebe HO, Tesfaye AA, Abtew WG.Association mapping in bambara groundnut [*Vigna subterranea* (L.) Verdc.] reveals loci associated with agro-morphological traits. BMC Genomics. 2023;**24**(1):593

[18] Ajilogba CF, Olanrewaju OS, and Babalola OO. Improving bambara groundnut production: Insight into the role of omics and beneficial bacteria. Frontiers in Plant Science. 2022;**13**:836133. DOI: 10.3389/fpls.2022.836133

[19] Temegne NC, Foh TD, Taffouo VD, Wakem GA, Youmbi E. Effect of mycorrhization and soluble phosphate on growth and phosphorus supply of Voandzou [*Vigna subterranea* (L.) Verdc.]. Legume Research-An International Journal. 2018;**41**(6): 879–884

[20] Ramolemana GM. The Phosphorus and Nitrogen Nutrition of Bambara Groundnut (*Vigna Subterranea* (L.) Verdc.) in Botswana Soils: An Exploratory Study. Wageningen University and Research; 1999:28225869

[21] Ibrahim II, Umar UM, Abubakar AI.Growth and yield attributes of groundnut (Arachis hypogea) as influenced by population density and phosphorous fertilizer rates on the Jos Plateau. Journal of Environmental Bioremediation and Toxicology. 2021;**4**(1):19–23

[22] Khan F, Siddique AB, Shabala S, Zhou M, Zhao C. Phosphorus plays key roles in regulating plants' physiological responses to abiotic stresses. Plants. 2023;**12**(15):2861

[23] Bitire TD, Abberton M, Tella EO, Edemodu A, Oyatomi O, and Babalola OO. Impact of nitrogen-fixation bacteria on nitrogen-fixation efficiency of Bambara groundnut [*Vigna subterranea* (L) Verdc] genotypes. Frontiers in Microbiology. 2023;**14**:1187250

[24] Kebede E. Contribution, utilization, and improvement of legumes-driven biological nitrogen

fixation in agricultural systems. Frontiers in Sustainable Food Systems. 2021;**5**:767998

[25] Kakraliya SK, Singh U, Bohra A, Choudhary KK, Kumar S, Meena RS, Jat ML. Nitrogen and legumes: A meta-analysis. Legumes for Soil Health and Sustainable Management. 2018; 277–314

[26] Zongo KF, Nandkangre H, Sanon A, Gouba M, Guebre D, Kambou DJ, Ouedraogo M. Improvement of Bambara groundnut productivity using organic manure fertilization doses on plinthite soil in sub-Saharan agrosystem of Burkina Faso. Nigeria Agricultural Journal. 2023;**54**(1):451–458

[27] El Bilali H, Kiebre Z, Nanema RK, Dan Guimbo I, Rokka VM, Gonnella M, . . . Acasto F. Mapping research on bambara groundnut (*Vigna subterranea* (L.) Verdc.) in Africa: Bibliometric, geographical, and topical perspectives. Agriculture. 2024;**14** (9):1541

[28] Soumare A, Diedhiou AG, Kane A. Bambara groundnut: A neglected and underutilized climate-resilient crop with great potential to alleviate food insecurity in sub-Saharan Africa. Journal of Crop Improvement [Internet]. 2022; **36**(5):747–767. DOI:10.1080/15427528.2021.2000908

[29] Erukainure OL, J O. Bambara–wheat composite flour: Rheological behavior of dough and functionality in brea. Food Science & Nutrition. 2016;**4** (6):852–857

[30] Mubaiwa J, Fogliano V, Chidewe C, Linnemann AR. Bambara groundnut (*Vigna subterranea* (L.) Verdc.) flour: A functional ingredient to favour the use of an unexploited sustainable protein source. PloS One. 2018;**13**(10):e0205776

[31] Kapembe MNP. The Effect of Different Processing Methods on the Nutritional, Anti-nutritional and Emulsifying Properties of Inoculated and Noinoculated Bambara Groundnut (*Vigna Subterranea*) Cultivars [Master thesis]. Windhoek: University of Namibia; 2024

[32] Mbuma NW, Labuschagne M, Siwale J, and Hugo A. Diversity in seed protein content, selected minerals, oil content and fatty acid composition of the Southern African Bambara groundnut germplasm collection. Journal of Food Composition and Analysis. 2022;**109**:104477

[33] Traoré-Barro A, Thiombiano EA, Gnankambary K, Rabo M, Coulibaly O, Traoré S, Traoré O. Characterization of Seed Storage Proteins in Eight Bambara Groundnut Landraces in Burkina Faso. Agricultural Sciences. 2023;**14** (9):1268–1276

[34] Majola NG, Gerrano AS, Shimelis H, and Mashilo J. Analyses of phytochemicals and nutritional compositions in Bambara groundnut (*Vigna subterranea* [L.] Verdc). Journal of Agriculture and Food Research. 2025;**19**:101610

[35] Piccinin E, Cariello M, De Santis S, Ducheix S, Sabbà C, Ntambi JM, Moschetta A.Role of oleic acid in the gut-liver axis: From diet to the regulation of its synthesis via stearoyl-CoA desaturase 1 (SCD1). Nutrients. 2019;**11**(10):2283

[36] Lin Tan X, Azam-Ali S, Goh EV, Mustafa MA, Chai HH, Kuan Ho W, . . . Massawe F. Bambara groundnut: An

underutilized leguminous crop for global food security and nutrition. Frontiers in Nutrition. 2020;7:276

[37] Adeleke O, Adiamo OQ, Fawale OS, Olamiti G.Effect of soaking and boiling on anti-nutritional factors, oligosaccharide contents and protein digestibility of newly developed Bambara groundnut cultivars. Turkish Journal of Agriculture-Food Science and Technology. 2017;**5**(9):1006–1014

[38] Gqaleni P *Nutritional Value of Bambara Groundnut (Vigna Subterranea (L.) Verdc.): a Human and Animal Perspective*. (Doctoral dissertation, University of KwaZulu-Natal, Pietermaritzburg). 2014

[39] Unigwe AE, Doria E, Adebola P, Gerrano AS, Pillay M. Anti-nutrient analysis of 30 Bambara groundnut (*Vigna subterranea*) accessions in South Africa. Journal of Crop Improvement. 2018;**32**(2):208–224

[40] Mubaiwa J, Fogliano V, Chidewe C, Linnemann AR. Hard-to-cook phenomenon in bambara groundnut (*Vigna subterranea* (L.) Verdc.) processing: Options to improve its role in providing food security. Food Reviews International. 2017;**33**(2):167–194

[41] Fwanyanga FM, Horn LN, Sibanda T, and Reinhold-Hurek B. Prospects of rhizobial inoculant technology on Bambara groundnut crop production and growth. Frontiers in Agronomy. 2022;**4**:1004771

[42] Aviara NA, Lawal AA, Atiku AA, Haque MA.Bambara groundnut processing, storage and utilization in north eastern Nigeria. Continental Journal of Engineering Sciences. 2013;**8**(1):28–36

[43] Nkhata SG, Ayua E, Kamau EH, Shingiro JB.Fermentation and germination improve nutritional value of cereals and legumes through activation of endogenous enzymes. Food Science & Nutrition. 2018;**6**(8):2446–2458

[44] Knez E, Kadac-Czapska K, Grembecka M.Effect of fermentation on the nutritional quality of the selected vegetables and legumes and their health effects. Life. 2023;**13**(3):655

[45] Jideani VA, Jideani AI, Jideani VA, Jideani AI. Nutritional composition of Bambara groundnut. Bambara Groundnut: Utilization and Future Prospects. 2021;75–85

[46] Valombola JS, Akundabweni LM, Awala SKSK, Hove K. Agronomic and morphological diversity of Bambara groundnut (*Vigna subterranea* (L.) Verdc.) accessions in North-Central Namibia. Welwitschia International Journal of Agricultural Sciences. 2019;**1**(1):88–99. http://journals.unam.edu.na/index.php/WIJAS/article/view/1369/1173

[47] Amarteifio JO, Tibe O, Njogu RM. The Nutrient Composition of Bambara Groundnut Landraces (Vigna subterreanea,(L.) Verdc.) Cultivated in Southern Africa. 2010

[48] Fleißner KWE *The Improvement of Bambara Groundnut Production in Northern Namibia by Means of Breeding Strategies and Agronomic Investigations*. (Doctoral dissertation, Technische Universität München). 2006

[49] Hilukwa R, Franke AC, Labuschagne M, Wanga MA, Hukununa RMK, Hangula MN, Sarsu F. Radio-sensitivity of selected namibian landrace groundnut (Arachis hypogaea

L.) genotypes to gamma radiation. Journal of Environmental Radioactivity. 2025;**282**:107618

[50] Kolech SA, De Jong W, Perry K, Halseth D, Mengistu F.Participatory variety selection: A tool to understand farmers' potato variety selection criteria. Open Agriculture. 2017;**2**(1):453–463

[51] Horn LN, Ghebrehiwot HM, Sarsu F, and Shimelis HA. Participatory varietal selection among elite cowpea genotypes in northern Namibia. Legume Research: An International Journal. 2017;**40**(6):995–1003

[52] Oseghale ES, Ezedinma FOC, Uguru MI. Effects of preparatory cultivation and porosity of growth medium on growth, reproductive efficiency and yield of Bambara groundnut (Voandzeia subterranean Thouars.). Nigerian Agricultural Journal. 1991;**25**:59–64

Chapter 5

Quinoa *(Chenopodium Quinoa)*: Biochemical Composition and Therapeutic Potential as a Functional Pseudocereal

Ayşe Usta

Abstract

Due to its unique biochemical structure and various bioactive components, quinoa (Chenopodium quinoa Willd.) is known as a nutrient-rich pseudograin with significant medicinal potential. Quinoa offers a wide range of phytochemicals, including flavonoids, phenolic acids, betalains, and saponins, in addition to high-quality proteins that comprise important minerals, vitamins, unsaturated fatty acids, all essential amino acids, and dietary fiber. These substances support its strong anti-inflammatory and antioxidant qualities. According to experimental data, quinoa-derived bioactives decrease lipid peroxidation, improve endogenous antioxidant defense systems by modifying the Nrf2 signaling pathway, and scavenge reactive oxygen species (ROS). Additionally, quinoa consumption has been linked to the suppression of pro-inflammatory cytokines such as TNF-α, IL-1β, and IL-6, partly due to its inhibition of NF-κB activation. Quinoa has positive metabolic effects in addition to its anti-inflammatory and antioxidative properties. Its low glycemic index and high fiber content contribute to improved glycemic control and insulin sensitivity. Consuming quinoa has also been associated with positive changes in lipid metabolism, such as lower levels of triglycerides, total cholesterol, and LDL cholesterol. Potential preventive roles against cardiovascular illnesses, non-alcoholic fatty liver disease, type 2 diabetes, and metabolic syndrome are suggested by emerging in vivo and clinical data. Quinoa is positioned as a promising functional food with therapeutic implications. More rigorously organized clinical trials are needed to better understand ideal intake levels, bioavailability of bioactive chemicals, and long-term health outcomes.

Keywords: quinoa, *Chenopodium quinoa*, bioactive compounds, antioxidant activity, anti-inflammatory effects, metabolic regulation

1. Introduction

1.1 The concept of pseudocereal

Technically, quinoa is not a true cereal; rather, it is a starchy dicotyledonous seed [1]. It does not belong to the Gramineae (Poaceae) family, which includes cereals such as wheat, rye, and barley [2]. Instead, quinoa is a broad-leaved plant belonging to the same family as spinach and sugar beet (*Amaranthaceae or Chenopodiaceae*) [3].

1.2 Historical use of quinoa

Quinoa (*Chenopodium quinoa Willd.*) is an annual plant belonging to the Amaranthaceae family and is native to the Andean region of South America [4]. Referred to as the "mother grain" by the Incas, quinoa has been cultivated for more than 7,000 years [5]. It was introduced to the European continent in the early 1980s. Owing to its high genetic diversity and its ability to grow at altitudes of up to 4,000 meters above sea level, quinoa is considered a highly adaptable crop in terms of agricultural production [6].

1.3 Importance of quinoa

Due to its remarkable tolerance to extreme climatic and soil conditions, such as drought, frost, and high salinity, quinoa is considered a strategic crop for combating global climate change and ensuring food security [7]. Following the UN's declaration of 2013 as the "International Year of Quinoa," its popularity grew dramatically throughout the world [8]. Quinoa's so-called grain status makes it critical to modern dietary habits and food security [9].

Its high dietary fiber content, low glycemic index, and the presence of some bioactive compounds support blood sugar regulation [4]. While preventing obesity and metabolic syndrome [10], its anti-inflammatory potential helps prevent chronic inflammatory diseases [11]. In addition, quinoa has the capacity to both directly scavenge free radicals and strengthen the body's natural antioxidant defense systems [12].

Additionally, quinoa is an important food source and a vital raw material for the food packaging, pharmaceutical, and cosmetic industries due to its varied antibacterial properties [3]. According to the Food and Agriculture Organization of the United Nations (FAO), quinoa is one of the few plants that can supply every vital nutrient needed by the human body on its own [13].

1.4 Botanical and agronomic morphological characteristics of *Chenopodium quinoa*

Quinoa (*Chenopodium quinoa Willd.*), a member of the *Chenopodiaceae* family, is physiologically classified among C3 plants [14]. The plant has a well-developed and branched taproot system that provides tolerance to drought conditions. Its erect stem can grow to a height of approximately 40–150 cm [15]. The leaves are broad and thick and are attached to somewhat woody stems. They typically exhibit serrated margins with patterns resembling a goosefoot and are generally triangular in shape. In young plants, the leaves are usually green; however, as the plant matures, they might take on various hues, such as purple, crimson, or yellow. The flowers form in panicle-like clusters, and the flowering period typically occurs between July and

August. The flowers are hermaphroditic and predominantly self-pollinated [16]. The seeds, which develop in clusters along the panicles, are generally round and measure approximately 2–3 mm in diameter. Although the grain weight varies depending on the cultivar, it typically ranges between 1.99 g and 5.08 g [17].

2. Nutritional composition of quinoa

Quinoa possesses a superior nutritional profile compared to conventional cereals. Seeds generally contain 64% carbohydrates, 17% protein, 6% fat, and about 4% dietary fiber [18]. The nutritional composition of raw black, yellow, and red quinoa flours shows significant variations in terms of protein quality, fiber content, mineral richness, and the phytochemicals they contain (**Table 1**) [19].

2.1 Protein composition

The protein content of quinoa varies between 12 and 23%, depending on genotypic and environmental factors [20], which is higher than that of wheat, maize, and rice [21]. One of its most important characteristics is that it contains all nine essential amino acids that cannot be synthesized by the human body – namely phenylalanine, methionine, histidine (which plays a key role during childhood), isoleucine, valine, leucine, lysine, threonine, and tryptophan – in a well-balanced proportion, particularly lysine and arginine [22]. All of these essential amino acids, which are crucial for the development and maintenance of human physiological functions, are present in quinoa seeds. Since quinoa contains relatively high levels of amino acids that are typically limited in other cereals – such as cysteine, lysine, and methionine – it is considered an

Parameter	Black	Yellow	Red
Moisture (g/100 g)	11.53 ± 0.07^{b}	12.04 ± 0.12^{a}	10.90 ± 0.13^{b}
Ash (g/100 g)	2.57 ± 0.10^{a}	2.32 ± 0.11^{a}	2.58 ± 0.74^{a}
Lipid (g/100 g)	5.15 ± 0.14^{a}	3.89 ± 0.28^{a}	4.86 ± 0.68^{a}
Protein (g/100 g)	20.90 ± 0.17^{a}	19.34 ± 0.52^{b}	20.25 ± 0.20^{a}
TDF (% w/w)	22.97 ± 0.20^{a}	19.43 ± 0.45^{b}	16.15 ± 0.90^{c}
IDF (% w/w)	22.65 ± 0.10^{a}	16.83 ± 0.14^{b}	14.57 ± 0.06^{c}
SDF (% w/w)	0.32 ± 0.06^{c}	2.40 ± 0.31^{a}	1.58 ± 0.80^{b}
D-glucose (g/100 g)	56.07 ± .45^{b}	64.36 ± 2.63^{a}	66.4 ± 2.63^{a}
D-fructose (g/100 g)	4.04 ± 0.71^{a}	2.55 ± 0.63^{b}	1.70 ± 0.30^{c}
Total carbohydrates (g/100 g)	60.12 ± 3.16^{c}	66.9 ± 3.26^{a}	68.1 ± 2.93^{b}
RS (g/100 g)	9.05 ± 0.98^{b}	8.90 ± 0.75^{b}	10.29 ± 0.79^{a}
NRS (g/100 g)	44.36 ± 2.62^{b}	45.90 ± 3.07^{a}	44.28 ± 3.07^{b}
TS (g/100 g)	53.41 ± 3.6^{b}	54.8 ± 3.75^{a}	54.57 ± 3.79^{a}

Abbreviations: IDF, Insoluble dietary fiber; NRS, Non-resistant starch; RS, Resistant starch; SDF, Soluble dietary fiber; TDF, Total dietary fiber; TS, Total starch.

Table 1.
Raw red, black, and yellow quinoa flours' nutritional makeup based on dry weight.

excellent source of protein [23]. While quinoa competes with wheat in terms of protein quantity, it surpasses most cereals in protein quality [13]. Among quinoa varieties, black quinoa exhibits the highest protein content at 20.90%, followed by red quinoa at 20.25% and yellow quinoa at 19.34% [19].

Quinoa proteins also demonstrate high digestibility and provide a biological value comparable to that of milk protein (casein) [24, 25]. The biological value (BV) of quinoa protein is approximately 73%, which is similar to that of beef (74%) and significantly higher than that of cereals such as wheat (49%), rice (56%), and maize (36%) [26]. In vitro studies have shown that the protein digestibility of quinoa exceeds 76.9%, and this value can increase to as high as 86% in fermented quinoa flour [27].

2.2 Lipid composition

The lipid content of quinoa varies between 1.8% and 9.5%, depending on genotypic characteristics, growing medium, and the analytical method used. Quinoa contains a higher lipid content compared to common cereals such as wheat (2.5%) and rice (1.3%) [28]. Notably, quinoa is free of cholesterol [29]. Although there are no substantial differences among quinoa varieties, the lipid content generally ranges between 3.8% and 5.15% [19].

The lipid fraction of quinoa is predominantly composed of polyunsaturated fatty acids (PUFAs), particularly those beneficial for cardiovascular health, such as omega-6 (linoleic acid), accounting for approximately 45–55%, and omega-3 (α-linolenic acid), accounting for about 5–10%. Together, these fatty acids constitute roughly 50–60% of the total fatty acid composition [30, 31].

2.3 Carbohydrate composition

The carbohydrate content of quinoa varies widely, ranging from 53% to 77% on a dry matter basis, depending on genotypic characteristics, growing conditions, and the analytical method used [32]. Starch, which makes up the largest portion of carbohydrates, generally accounts for between 52.2% and 73.4% [33]. Among quinoa varieties, red quinoa is slightly richer in resistant starch (10.29 g/100 g) compared with others [19]. Amylopectin, which constitutes the majority of quinoa starch, is readily hydrolyzed and absorbed in the small intestine. In contrast, resistant starch fractions, such as amylose, along with dietary fibers, escape digestion in the small intestine and reach the large intestine [34]. There, they are fermented by the intestinal microbiota to produce short-chain fatty acids (SCFAs), which are subsequently absorbed through the intestinal wall and serve as an energy source for the host [35].

2.4 Fiber and glycemic index content

The total dietary fiber content of quinoa varies widely, ranging from 7% to 23%, depending on genotypic characteristics (such as color variations) and growing conditions [36]. Quinoa is also rich in dietary fiber, which supports digestive health. Depending on the variety, quinoa contains significantly higher dietary fiber than rice (0.2–2.8%) and corn (1.4–7.3%) [37]. Approximately 80% of the fiber is insoluble, while 20% is soluble [38]. Among the varieties, black quinoa is the richest in both total dietary fiber (22.97%) and insoluble fiber (22.65%). Quinoa has a higher fiber quality and soluble fiber content than traditional grains [39]. Red quinoa has 16.15%

total fiber, while yellow quinoa has 19.43% [20]. Due to its low D-glucose concentration, black quinoa is also the option with the lowest glycemic index [40]. **Table 2** shows a comparison of the nutritional composition of quinoa seeds and some commonly used cereals (g/100 g, on a dry matter basis) [41].

2.5 Vitamin content

Quinoa is notably richer than wheat and rice in B-group vitamins, particularly riboflavin (B2) and folic acid (B9) [42]. It also contains higher levels of vitamin E (tocopherols), which contribute to lipid stabilization [43]. Quinoa oil is especially abundant in γ-tocopherol, providing protection against the oxidation of fatty acids [44]. **Table 3**: Vitamin content of uncooked quinoa (per 100 g) [45].

2.6 Mineral content

Quinoa is particularly rich in minerals such as calcium, magnesium, iron, and zinc, with levels significantly higher than those found in wheat, maize, and rice [1, 46]. The ash content, which represents the total mineral content, is similar among quinoa varieties, ranging from 2.32% to 2.58% [19]. Red quinoa is richer than the other two varieties in essential minerals such as potassium (6,322 ppm), iron (41.4 ppm), sodium (294.5 ppm), and zinc (35 ppm). In contrast, yellow quinoa shows a slight advantage over black and red varieties in calcium (478.7 ppm) and magnesium (1,550.3 ppm). These data indicate that black quinoa stands out for its high protein and fiber content, whereas red quinoa excels in mineral and antioxidant richness [19]. Quinoa generally contains minerals such as calcium, magnesium, potassium, and phosphorus in highly bioavailable forms [47]. **Table 4** shows the mineral composition of quinoa and barley samples (mg/100 g, on a dry weight basis) [30].

3. Bioactive compounds in quinoa

Quinoa is a rich reservoir of diverse phytochemicals and bioactive compounds that support metabolic, cardiovascular, and gastrointestinal health. These substances transform quinoa from a simple source of nutrients into a "functional food" [48].

	Carbohydrates	Protein	Fat	Fiber	Ash
Quinoa	69.0	16.5	6.3	3.8	3.8
Barley	80.7	10.8	1.9	2.2	4.4
Corn	81.1	10.2	4.7	11.7	2.3
Oats	69.8	11.6	5.2	2.9	10.4
Rice	80.4	7.6	2.2	3.4	6.4
Rye	80.1	13.4	1.8	2.1	2.6
Wheat	78.4	14.3	2.3	2.2	2.8

Table 2.
Contrasting the nutritional value of various grains with 100 grams of quinoa.

Vitamins	/100 g	Unit
Thiamine	0.360	mg
Riboflavin	0.318	mg
Niacin	1.520	mg
Pantothenic acid	0.772	mg
Vitamin B6	0.487	mg
Total folate	184	µg
Choline	70.2	mg
Betaine	630.4	mg
Vitamin A	14	IU
Vitamin A	1	µg
β-carotene	8	µg
β-cryptoxanthin	1	µg
Zeaxanthin + Lutein	163	µg
Vit E	2.44	mg
β-tocopherol	0.08	mg
γ-tocopherol	4.55	mg
λ-tocopherol	0.35	mg

Table 3.
Vitamin values of uncooked quinoa.

Minerals	Quinoa	Barley
Ca	86.3	29
Mg	502	79
K	732	280
P	411	221
Fe	15	2.5
Zn	4	2.1

Table 4.
Mineral composition of quinoa and barley samples.

3.1 Phenolic compounds

The most prevalent type of phytochemicals in quinoa is phenolic compounds, which are mostly found in the pericarp, or the seed coat's outer layers. Ferulic acid and vanillic acid, along with their derivatives, are the main phenolic acids [49]. Quinoa has at least 23 distinct phenolic chemicals, according to studies [50]. Seed color has a major impact on phenolic content; darker types, such as black and red quinoa, have higher phenolic contents and more potent antioxidant activity than white kinds [19]. These substances exist in bound, conjugated, and free forms [51]. Black quinoa has the highest anthocyanin content (19.3 mg CGE/L) among the varieties, while red quinoa has a higher total phenolic content (338.9 mg/100 g) [24].

Intestinal bacteria convert phenolic substances into physiologically active metabolites, which are then absorbed through the colon wall [52].

3.2 Flavonoids

Kaempferol, quercetin, and their derivatives make up the majority of quinoa's flavonoid profile [50]. Quinoa helps reduce inflammation and shields cells from oxidative stress through flavonoids and polyphenols [53].

3.3 Saponins

Quinoa gets its bitter flavor from saponins, which are found in the outer seed coat [54]. To remove the bitterness, the seeds are usually washed before consumption [55]. In different sections of the quinoa plant, about 30 distinct saponins have been identified [56]. One of quinoa's most distinctive secondary metabolites, saponins reside in the outer layer of the seed and provide defense against microbial attack [57].

Quinoa seeds can be black, yellow, or red, with color variation influenced by the country of cultivation and corresponding differences in saponin and other bioactive compound content [58]. Red and yellow quinoa varieties contain higher levels of saponins (95–96 mg/g) and phytic acid (2.1–2.2 g/100 g) compared to black quinoa [19].

3.4 Phytosterols

Quinoa seeds contain phytosterols, lipophilic compounds structurally similar to cholesterol [59]. By competing with cholesterol for absorption in the intestine, they help reduce total cholesterol levels in the body [60]. The most abundant phytosterol in quinoa is β-sitosterol, followed by campesterol and stigmasterol [30].

3.5 Phytoecdysteroids

Phytoecdysteroids are produced by plants as a defense mechanism against insects [61]. Among edible plants, quinoa ranks highest in phytoecdysteroid content, ranging from 138 to 570 μg/g [62]. The most common compound in this group is 20-hydroxyecdysone (20HE), which has potential benefits in the management of metabolic syndrome and diabetes [63].

3.6 Betalains

Betalains are nitrogen-containing pigments responsible for the yellow, red, and black colors of quinoa seeds [64]. They are classified into two main types: Betacyanins, which provide red-violet coloration (particularly betanin and isobetanin), and betaxanthins, which give yellow-orange hues. Betacyanins present in red and black quinoa varieties exhibit strong free radical-scavenging activity [65].

4. Therapeutic and metabolic effects

Quinoa contains numerous phytochemicals, such as saponins, that contribute to its health-promoting properties. Evidence indicates that quinoa exhibits strong antioxidant, anti-inflammatory, antidiabetic, cardioprotective, anticancer, antimicrobial, and hepatoprotective effects, and may also provide benefits in the management of metabolic syndrome and celiac disease [66, 67].

4.1 Antioxidant effect

Quinoa contains a wide array of bioactive compounds that neutralize free radicals in the body, protecting cells against oxidative stress and helping to reduce the risk of chronic diseases [66]. This property is largely attributed to its rich phytochemical content [68]. Flavonoids, such as quercetin and kaempferol, are abundant in quinoa seeds and leaves, exhibiting strong capacities to inhibit lipid oxidation and scavenge free radicals [69]. The pigments responsible for the red and black coloration of quinoa varieties, namely betanin and isobetanin, also display potent radical-scavenging activity. Some studies suggest that the antioxidant capacity of betalains may even surpass that of polyphenols [49].

Saponins, often removed due to their bitter taste, have been shown to possess anti-inflammatory and antioxidant activities [57]. Polysaccharides derived from quinoa seeds and bioactive peptides produced during enzymatic hydrolysis or quinoa protein digestion exhibit strong antioxidant activity [70, 71]. Vitamins C and E, which protect cell membranes and prevent lipid oxidation, are abundant in quinoa oil [72].

Consuming quinoa boosts the quantity of endogenous antioxidant enzymes such as superoxide dismutase (SOD), catalase (CAT), and glutathione peroxidase (GSH-Px). It has been demonstrated that quinoa proteins and hydrolyzates enhance cellular antioxidant defense mechanisms by activating the Nrf2/HO-1 signaling pathway [73]. Quinoa extracts are highly effective at scavenging superoxide, hydroxyl, ABTS, and DPPH anion radicals [74]. Compared to yellow quinoa, dark-colored variants like black and red quinoa have higher phenolic contents and more potent antioxidant activity [50]. Furthermore, the antioxidant capacity of quinoa roots, sprouts, and leaves may be greater than that of the seeds [75].

Quinoa also modulates energy metabolism, particularly through protein hydrolyzates and peptides, by activating the AMP-activated protein kinase (AMPK) pathway. In studies with obese mice, dietary supplementation with quinoa seeds increased hepatic AMPK mRNA expression [76]. Quinoa has been reported to improve metabolic parameters by modulating AMPK/PPAR-γ signaling pathways. AMPK activation supports fat oxidation, enhances exercise endurance, regulates hepatic glucose production (gluconeogenesis), and increases glucose uptake in skeletal muscle, contributing to overall glucose homeostasis. Additionally, AMPK activation reduces hepatic fatty acid synthesis while promoting lipid oxidation, thereby maintaining metabolic balance [77].

4.2 Anti-inflammatory effect

Quinoa contains a wide range of bioactive compounds that modulate inflammatory processes in the body and contribute to lowering the likelihood of chronic diseases. Oxidative stress and chronic inflammation, which are among the primary causes of endothelial dysfunction, may be suppressed through quinoa consumption [66]. Saponins present in quinoa, particularly monodesmosidic forms, protect vascular tissues against damage by reducing inflammatory cytokines such as nitric oxide (NO), TNF-α, and IL-6, as well as decreasing oxidative stress markers such as malondialdehyde (MDA) [78].

Chenopodin, the major protein found in quinoa, and peptides such as lunasin, possess the ability to inhibit inflammatory markers [79]. Quercetin and kaempferol exhibit anti-inflammatory activity by influencing cellular signaling pathways and metabolic processes. It has also been reported that red and black quinoa varieties possess higher anti-inflammatory potential than white varieties [80].

Quinoa oil, along with its omega-3 and omega-6 fatty acids and vitamin E (tocopherol), may help reduce inflammation by supporting gene regulation and scavenging free radicals [44]. Quinoa consumption also regulates inflammatory responses by inhibiting the NF-κB and TLR4 signaling pathways, which play key roles in the initiation of inflammation [81].

Quinoa extracts have been shown to reduce hepatic inflammation and oxidative stress in non-alcoholic fatty liver disease (NAFLD) [82]. In a study conducted on overweight postmenopausal women, the consumption of quinoa flakes for four weeks was found to reduce IL-6 levels [83]. Moreover, it has been reported that quinoa protein intervention suppresses exercise-induced inflammation and TNF-α levels following high-intensity exercise, supporting muscle recovery [73].

Cyclooxygenase-2 (COX-2) is one of the key enzymes involved in triggering inflammatory processes in the body. Extractable polyunsaturated fatty acids (PUFAs) and phenolic compounds present in cooked quinoa have been shown to significantly downregulate COX-2 mRNA expression in a dose-dependent manner. The bioactive components of quinoa reduce the severity of inflammatory responses by suppressing this enzyme, as well as other inflammatory markers such as IL-6, IL-8, TNF-α, and IL-1β [49, 84].

Additionally, through its phytochemical constituents, quinoa may help regulate inflammation and suppress cancer development by modulating intracellular signaling networks, including the MAPK (mitogen-activated protein kinase) signaling pathways and regulatory mechanisms such as MAP kinase phosphatase-5 [85].

4.3 Antidiabetic effect

Quinoa is considered a low glycemic index (GI) food due to its slowly digestible starch structure and high dietary fiber content [40]. For example, the GI value of quinoa milk is 52, whereas that of rice milk is 79 [86]. Similarly, the glycemic index of quinoa (approximately 53) is considerably lower than that of rice (68) and wheat (70) [87]. Replacing a portion of wheat flour with quinoa flour increases the antioxidant activity, mineral content, and essential amino acid composition of bread while reducing its glycemic index [88]. This trait minimizes abrupt spikes in blood glucose levels, or postprandial hyperglycemia, by slowing the body's digestion of food [89].

Quinoa also helps regulate digestion and has a lower glycemic index compared with cereals such as rice, which contributes to improved glycemic control [90]. Therefore, consuming quinoa instead of white rice or refined flour-based products may represent an effective dietary strategy for the management and prevention of the progression of type 2 diabetes [91].

Small peptides released during the digestion or enzymatic hydrolysis of quinoa proteins have been shown to exhibit antihypertensive (ACE-inhibitory), antidiabetic, and anti-fatigue effects [92].

Slow glucose release reduces the insulin secretion burden on pancreatic β-cells, thereby contributing to the protection of these cells [13]. Phenolic compounds, bioactive peptides, and polysaccharides present in quinoa can inhibit α-amylase and α-glucosidase enzymes responsible for the conversion of starch into glucose, thereby delaying intestinal glucose absorption and limiting the rise in blood glucose levels [93].

The dipeptidyl peptidase-4 (DPP-IV), which breaks down incretin hormones that control insulin release, may also be inhibited by peptides generated from quinoa. By inhibiting DPP-IV, quinoa bioactive peptides may prolong the activity of incretin hormones that stimulate insulin secretion, thereby improving the regulation of insulin levels during the postprandial phase [94].

20-Hydroxyecdysone (20HE), which is abundant in quinoa, has been observed to significantly reduce fasting blood glucose levels, enhance insulin sensitivity, and alleviate obesity-related diabetic symptoms [50]. In particular, polyphenols from black quinoa have been reported to increase glucose utilization and glycogen synthesis by activating specific intracellular signaling pathways (IRS1/PI3K/Akt) associated with the improvement of insulin resistance [95].

Replacing refined rice or wheat products with quinoa has been reported to regulate lipid levels and blood glucose in individuals with a reduced ability to tolerate glucose. In addition, quinoa-based diets have been shown to reduce postprandial glycemic fluctuations in prediabetic elderly individuals [96]. Studies in diabetic mice have also demonstrated that quinoa intervention reduces fasting blood glucose levels, improves glucose tolerance, and enhances insulin levels [91].

4.4 Cardioprotective properties

Phytosterols present in quinoa, particularly β-sitosterol, help reduce total cholesterol and low-density lipoprotein (LDL, commonly referred to as "bad cholesterol") levels in the body by competing with cholesterol for absorption in the intestines [59, 97]. Furthermore, quinoa seeds' saponins and dietary fibers have been shown to have cholesterol-lowering properties, promoting cardiovascular health and cholesterol control [67].

Bioactive peptides found in quinoa function as angiotensin-converting enzyme (ACE) inhibitors and have antihypertensive effects via controlling blood pressure [92]. Quinoa protein hydrolyzates dramatically lower systolic blood pressure, according to research conducted on rats with spontaneous hypertension [98].

Metabolic syndrome is characterized by elevated triglycerides, poor HDL cholesterol, and elevated blood glucose. It has been shown that eating dishes enhanced with quinoa improves lipid profiles by lowering total cholesterol, triglycerides (TG), LDL, and very low-density lipoprotein (VLDL), while promoting increases in HDL [99, 100]. Quinoa's unsaturated fatty acids and phytosterols help decrease LDL cholesterol, which lowers the risk of atherosclerosis and indirectly supports endothelium integrity [101].

Because of its unique bioactive substances and antioxidant capability, quinoa also demonstrates preventive and restorative effects on endothelium function. Flavonoids, which are present in quinoa, improve blood flow while lowering the risk of blood cell aggregation, decreasing vascular permeability and fragility, and encouraging coronary vasodilation [102]. As a result, quinoa promotes endothelium health by reducing vascular inflammation, improving blood flow, and preserving vascular resistance [103].

4.5 Anticancer effect

Quinoa's ability to halt cell damage and inhibit the growth of cancer cells demonstrates its anti-cancer potential. Specifically, it has been observed that kaempferol significantly inhibits the growth of cancer cells [104]. By avoiding DNA damage and neutralizing reactive oxygen species (ROS), polyphenols and flavonoids reduce oxidative stress. This process aids in preventing DNA damage, which, in turn, prevents the development of cancer [105].

Quinoa has a lot of kaempferol, which activates the Nrf2 signaling pathway involved in cellular defense mechanisms. This results in the activation of cytoprotective genes and the maintenance of cellular redox balance, both of which help prevent the development of cancer [106].

By lowering inflammatory cytokines like TNF-α and triggering apoptosis, quinoa oil nanocapsules have been demonstrated to reduce tumor growth in breast cancer mice models. This action is linked to the reduction of proto-oncogenes, including MYC and PIK3CA, which encourages cancer cells to undergo apoptosis [107].

Peptides and saponins derived from the quinoa plant have also been shown to reduce cell viability and prevent proliferation in colon cancer cells [108]. These substances possess significant anticancer qualities, even though they are frequently eliminated during processing stages like washing [28].

Additionally, it has been demonstrated that some peptides obtained from quinoa protein hydrolyzates, specifically FHPFPR, NWFPLPR, and HYNPYFPG, inhibit the activity of HDAC1 (histone deacetylase 1), which modifies the expression of genes linked to cancer and inhibits the growth of cancer cells [108].

4.6 Antimicrobial effect

Quinoa has antiviral, antifungal, and antibacterial properties [109, 110]. The plant's inherent defensive mechanisms, which include bioactive peptides, phenolic compounds, and saponins, are mostly responsible for this antibacterial activity [3]. Saponins are thought to be the most powerful antibacterial constituents of quinoa, serving as a chemical barrier that prevents microorganisms from entering the plant [57].

Alkali-transformed quinoa saponins may find use in oral care products since they have been shown to have greater antibacterial activity than native saponins against three distinct bacterial species that cause halitosis (oral malodor). Furthermore, they have shown inhibitory actions against *Bacillus cereus*, *Staphylococcus aureus*, and *Staphylococcus epidermidis* [111].

The plant pathogen *Botrytis cinerea's* mycelial growth and spore germination are markedly inhibited by quinoa saponins treated with alkali [112]. Additionally, it has been discovered that crude quinoa saponins are more efficient than purified saponins at preventing the fungus *Candida albicans* from growing. This is probably because the bioactive components work in concert [107].

Saponins work by breaking down bacterial cell walls and disrupting plasma membranes, which causes internal components to seep out and eventually leads to cell death [110].

It has also been demonstrated that quinoa seed phenolic extracts have strong antibacterial action against common foodborne pathogens like *Staphylococcus aureus* and *Escherichia coli* [113]. Furthermore, even at low quantities, quinoa essential oils have shown inhibitory actions against fungal species such as *Mucor pusillus, Fusarium oxysporum*, and *Aspergillus niger* [114].

By inhibiting the growth of total bacteria, molds, and yeasts, modified quinoa proteins and quinoa peptide-loaded liposomes may function as natural preservatives in meat products like burgers [115]. Furthermore, when added to food packaging materials, complexes made of quinoa starch and gold nanoparticles have been shown to successfully stop the growth of S. aureus and E. coli, extending the shelf life of the product [116].

4.7 Hepatoprotective effect

In models of obesity and diabetes, quinoa eating has been demonstrated to dramatically lower hepatic cholesterol levels and lipid accumulation (steatosis) [117]. By reducing the levels of aspartate aminotransferase (AST) and alanine aminotransferase (ALT), which are clinical indicators of liver damage, it maintains liver function [91].

Quinoa improves glucose homeostasis and fatty acid oxidation by upregulating the production of PGC-1α and SIRT1 proteins in the liver [118]. Quinoa prevents cellular damage by modifying endoplasmic reticulum (ER) stress indicators in hepatic tissue, such as GRP78 and CHOP [119]. Moreover, it has been documented that red quinoa husk extracts reduce liver fibrosis by inhibiting the TGF-β1 signaling pathway [120]. Additionally, quinoa prevents lipid buildup via controlling fatty acid metabolism and downregulating HMG-CoA reductase mRNA levels in the liver [121].

4.8 Benefits for celiac disease and gluten intolerance

Quinoa's inherent gluten-free status is one of its greatest tactical benefits [1]. Quinoa is a great substitute for wheat for people with celiac disease or gluten intolerance because it doesn't contain gluten [122]. Quinoa is especially crucial for the creation of gluten-free products for people with celiac disease [123]. Research on gluten-free flours has shown that breads made with quinoa flour have better volume and texture than breads made with rice or maize flour [124].

4.9 Anti-obesity and metabolic syndrome effects

It has been demonstrated that quinoa-derived peptides reduce obesity by altering the liver's PPAR-α/γ signaling pathways [125]. Quinoa consumption has been shown in animal experiments to lower low-grade chronic inflammation linked to obesity, as evidenced by indicators like IL-6 and protein carbonyl levels [126]. All of these studies highlight quinoa's potential to help control diabetes, reduce the risk of obesity, and improve cardiovascular health by decreasing cholesterol [4, 121].

4.10 Modulation of gut microbiota

Quinoa's bound phenolics are closely linked to cell wall components and are essential for preserving colon health after digestion [127]. Quinoa-based diets have been observed to positively modulate the gut microbiota and restore the mucosal barrier, thereby reducing inflammation in conditions such as colitis [128]. Quinoa polysaccharides, which exhibit immunoregulatory and prebiotic properties, further promote a healthy gut microbial composition [129].

Consumption of quinoa beneficially alters the gut flora, increasing the production of short-chain fatty acids (SCFAs), which contribute to improved glucose metabolism [130]. The high dietary fiber content of quinoa slows digestion, enhances satiety, and limits glucose absorption from the intestines [37]. Moreover, quinoa intake supports SCFA production through gut microbiota modulation, which, in turn, promotes hormonal signaling pathways that further improve glucose metabolism [13].

5. Industrial applications

5.1 Washing

Mostly employed to extract saponins from the seeds' outer layer, washing eliminates the bitter taste of quinoa and improves its sensory quality. However, the removal of saponins can also lead to significant losses of minerals such as sodium, calcium, magnesium, and potassium, as well as dietary fiber and phenolic compounds [131].

5.2 Fermentation

Fermentation of quinoa with probiotic bacteria enhances its functional properties. It leads to a marked increase in protein content and in vitro protein digestibility [132]. The breakdown of phytates during fermentation facilitates mineral bioavailability [133]. Additionally, enrichment in B vitamins, such as folate and riboflavin, has been observed [134]. Fermentation is one of the most effective methods for enhancing quinoa's antioxidant potential and phenolic content, with reported increases ranging from 2- to 12-fold [135].

5.3 Thermal processing

Thermal treatments are effective in preserving amino acids, lipids, and polyphenols while increasing quinoa digestibility [136]. However, significant losses can occur, with up to 40% of minerals, such as potassium, leaching into water and polyphenols decreasing by as much as 88% [137]. Partial losses in antioxidant chemicals can result from boiling or grinding [138].

5.4 Extrusion

Extrusion is a quick processing technique that uses high pressure and temperature. It improves digestibility by facilitating the denaturation of proteins and the gelatinization of starch [139]. However, it might lower the amounts of lipids, free polyphenols, and heat-sensitive vitamins [140]. While pre-washing is advised to eliminate bitterness, boiling is encouraged to maintain optimum nutritional content [141].

5.5 Technological processing and industrial applications

In the food sector, quinoa is a versatile ingredient. It is ground into flour and used to make pasta, bread, biscuits, and snacks [49]. Additionally, it is fermented to make drinks and plant-based milks [142]. Quinoa protein and starch have potential applications in cosmetic formulations, medicinal carrier systems, and eco-friendly biodegradable packaging sheets [143].

6. Conclusion and future perspectives

The majority of current clinical research has been carried out over brief periods of time and with small sample sizes. Therefore, more extensive and long-term research is required to validate quinoa's long-term benefits. It is still unclear what molecular processes underlie the bioavailability and activity of bioactive substances in the human body. It is more difficult to standardize data since different quinoa varieties (white, black, red) and processing techniques drastically change the bioactive composition. Although quinoa has shown therapeutic promise in cell culture and animal models through mechanisms such as AMPK activation, the Nrf2/HO-1 pathway, and NF-κB suppression, the direct relevance of these molecular effects to human physiology has not yet been fully demonstrated. In conclusion, more consistent clinical data are needed before quinoa's potent anti-inflammatory and anti-diabetic effects can be incorporated into established treatment approaches for people. Overall, quinoa's remarkable nutritional composition and variety of bioactive compounds make it stand out as a promising functional pseudocereal. Numerous biological processes, including cardioprotective, antioxidant, hepatoprotective, antidiabetic, and anti-inflammatory effects, are facilitated by the availability of vitamins, dietary fiber, unsaturated fats, high-quality proteins, minerals, and phytochemicals. Quinoa's reputation as a "superfood" is supported by its agronomic resilience and nutritional richness. It is also extremely helpful for people with celiac disease or gluten intolerance because it is free of gluten.

The majority of current clinical research has been carried out over brief periods of time and with small sample sizes. Therefore, more extensive and long-term research is required to validate quinoa's long-term benefits. It is still unclear what molecular processes underlie the bioavailability and activity of bioactive substances in the human body. It is more difficult to standardize data since different quinoa varieties (black, red, white) and processing techniques drastically change the bioactive composition. Although quinoa has been shown to have therapeutic potential in cell culture and animal models through pathways like AMPK activation, the Nrf2/HO-1 pathway, and NF-κB suppression, it is still unclear how directly these molecular effects relate to human physiology. In conclusion, more consistent clinical data are needed before quinoa's potent anti-inflammatory and anti-diabetic effects can be incorporated into established treatment approaches for people. Well-designed human studies are still required to define ideal intake amounts, the bioavailability of bioactive compounds, and long-term health consequences, despite growing experimental and clinical data supporting its health advantages. Although quinoa's commercial applications are currently limited by high production costs, advancements in processing technology and rising consumer awareness are anticipated to significantly expand quinoa's role in the contemporary diet.

To fully realize the promise of this resilient crop, future research should also focus on developing functional foods, dietary supplements based on quinoa, and sustainable farming methods.

Acknowledgment

The author acknowledges the use of notebooklm for language polishing of the manuscript.

Conflict of Interest

No conflicts of interest have been disclosed by the author.

Author details

Ayşe Usta
Faculty of Science, Department of Chemistry, Van Yuzuncu Yil University, Van, Türkiye

*Address all correspondence to: ayseusta@yyu.edu.tr

References

[1] Filho AM, Pirozi MR, Borges JT, Pinheiro Sant'Ana HM, Chaves JB, Coimbra JS. Quinoa: Nutritional, functional, and antinutritional aspects. Critical Reviews in Food Science and Nutrition. 2017;**57**(8):1618–1630. DOI: 10.1080/10408398.2014.1001811

[2] Navruz-Varli S, Sanlier N. Nutritional and health benefits of quinoa (Chenopodium quinoa Willd.). Journal of Cereal Science. 2016;**69**:371–376. DOI: 10.1016/j.jcs.2016.05.004

[3] Casalvara RFA, Ferreira BMR, Gonçalves JE, Yamaguchi NU, Bracht A, Bracht L, Comar JF, de Sá-nakanishi AB, de Souza CGM, Castoldi R, et al. Biotechnological, Nutritional, and therapeutic applications of Quinoa (Chenopodium quinoa Willd.) and its by-products: A Review of the Past Five-Year Findings. Nutrients. 2024;**16**:840

[4] Ahmad J, Khan I, Manzoor A, Abdullah Khan F, Al-Habsi N, Rahman MS. Quinoa: An underutilized pseudocereal with promising health and industrial benefits. Journal of Agricultural and Food Chemistry. 2025;**73**:14722–14741. DOI: 10.1021/acs.jafc.5c02291

[5] Connolly D, Minj J, Murphy KM, Solverson PM, Rust BM, Carbonero F. Impact of quinoa and food processing on gastrointestinal health: A narrative review. Critical Reviews in Food Science and Nutrition. 2025;**65**(28):5520–5533. DOI: 10.1080/10408398.2024.2416476

[6] Jacobsen SE. The worldwide potential for Quinoa (Chenopodium quinoa Willd.). Food Reviews International. 2025;**x19**(1-2):167–177. DOI: 10.1081/FRI-120018883

[7] López-Marqués RL, Nørrevang AF, Ache P, Moog M, Visintainer D, Wendt T, Østerberg JT, Dockter C, Jørgensen ME, Salvador AT, et al. Prospects for the accelerated improvement of the resilient crop quinoa. Journal of Experimental Botany. 2020;**71**(18):5333–5347. DOI: 10.1093/jxb/eraa285

[8] Bazile D, Bertero HD, Nieto C. State of the art report on quinoa around the world in 2013. FAO; 2015

[9] Alandia G, Rodriguez JP, Jacobsen SE, Bazile D, Condori B. Global expansion of quinoa and challenges for the Andean region. Global Food Security. 2015;**26**:100429. DOI: 10.1016/j.gfs.2020.100429

[10] Sezgin AC, Sanlier N. A new generation plant for the conventional cuisine: Quinoa (Chenopodium quinoa Willd). Trends in Food Science and Technology. 2019;**86**:51–58. DOI: 10.1016/j.tifs.2019.02.039

[11] Navarro-Perez D, Radcliffe J, Tierney A, Jois M. Quinoa seed lowers serum triglycerides in overweight and obese subjects: A dose-response randomized controlled clinical trial. Current Developments in Nutrition. 2017;**1**(9):e001321. DOI: 10.3945/cdn.117.001321

[12] Pereira E, Encina-Zelada C, Barros L, Gonzales-Barron U, Cadavez V, FerreiraI CFR. Chemical and nutritional characterization of Chenopodium quinoa Willd
T (quinoa) grains: A good alternative to nutritious food. Food Chemistry. 2019;**280**:110–114. DOI: 10.1016/j.foodchem.2018.12.068

[13] Zhang H, Li R. Effects of quinoa on cardiovascular disease and diabetes: A review. Frontiers in Nutrition. 2024;**11**:1470834. DOI: 10.3389/fnut.2024.1470834

[14] Geren H, Güre E. A Preliminary Study on the Effect of Different N and P Levels on the Grain Yield and Other Yield Components of Quinoa (Chenopodium quinoa Willd.). Ege University Ziraat Fakültesi Dergisi. 2017;**54**(1):1–8. DOI: 10.20289/zfdergi.297899

[15] Bhargava A, Shukla S, Ohri D. Genetic variability and interrelationship among various morphological and quality traits in quinoa (Chenopodium quinoa Willd.). Field Crops Research. 2007;**101**(1):104–116. DOI: 10.1016/j.fcr.2006.10.001

[16] Gordillo-Bastidas E, Diaz-Rizzolo DA, Roura E, Massanés T, Gomis R. Quinoa (Chenopodium quinoa Willd). from nutritional value to potential health benefits: An integrative review. Journal of Nutrition and Food Science. 2016;**6**(3):1–10. DOI: 10.4172/2155-9600.1000497

[17] Vilche C, Gely M, Santalla E. Physical properties of quinoa seeds. Biosystems Engineering. 2003;**86**(1):59–65. DOI: 10.1016/S1537-5110(03)00114-4

[18] Jancurová M, Minarovičová L, Dandár A. Quinoa – A review. Czech Journal of Food Sciences. 2009;**27**(2):71–79. DOI: 10.17221/32/2008-CJFS

[19] Manzanilla-Valdez ML, Boesch C, Orfila C, Montãno S, Hernández-alvarez AJ. Unveiling the nutritional spectrum: A comprehensive analysis of protein quality and antinutritional factors in three varieties of quinoa (Chenopodium quinoa Wild). Food Chemistry: X. 2024;**24**:101814. DOI: 10.1016/j.fochx.2024.101814

[20] Wang X, Zhao R, Yuan W. Composition and secondary structure of proteins isolated from six different quinoa varieties from China. Journal of Cereal Science. 2020;**95**:103036. DOI: 10.1016/j.jcs.2020.103036

[21] Hussain MI, Farooq M, Syed QA, Ishaq A, Al-Ghamdi AA, Hatamleh AA. Botany, nutritional value, phytochemical composition and biological activities of quinoa. Plants. 2021;**10**(11):2258. DOI: 10.3390/plants10112258

[22] Mariotti F, Gardner CD. Dietary protein and amino acids in vegetarian diets-a review. Nutrients. 2019;**11**:2661. DOI: 10.3390/nu11112661

[23] Repo-Carrasco-Valencia RAM, Serna LA. Quinoa (Chenepodium quinoa Willd) as a source of dietary fiber and other functional components. Ciencia Y Tecnología de Alimentos. 2011;**31**(1):225–230. DOI: 10.1590/S0101-20612011000100035

[24] Song H, Fu Q, Huang K, Zou Z, Chen L, Chen H, Ge S, Wang J, Guan X. Digestion characteristics of quinoa, Barley and mungbean proteins and the effects of their simulated gastrointestinal digests on CCK secretion in enteroendo-crine STC-1 cells. Food Function. 2022;**13**(11):6233–6243. DOI: 10.1039/d2fo00243d

[25] Comino I, de Lourdes Moreno M, Real A, Rodríguez-Herrera A, Barro F, Sousa C. The gluten-free diet: Testing alternative cereals tolerated by celiac patients. Nutrients. 2013;**5**(10):4250–4268. DOI: 10.3390/NU5104250

[26] Ng CY, Wang M. The functional ingredients of quinoa (Chenopodium quinoa) and physiological effects of consuming quinoa: A review. Food Frontiers. 2021;**2**(3):329–356. DOI: 10.1002/fft2.109

[27] Lorusso A, Verni M, Montemurro M, Coda R, Gobbetti M, C. G R. Use of fermented quinoa flour for pasta mak-ing and evaluation of the technological and nutritional features. LWT. 2017;**78**:215–221. DOI: 10.1016/j.lwt.2016.12.046

[28] Liu C, Ma R, Tian Y. An overview of the nutritionalprofile, processing technologies, and health benefits of quinoa with an emphasis on impacts of processing. Critical Reviews in Food Science and Nutrition. 2024;**64**(16):5533–5550. DOI: 10.1080/10408398.2022.2155796

[29] Miranda M, Vega-Galvez A, Quispe-Fuentes I, Rodriguez MJ, E. A MM. Nutritional aspects of six quinoa (Chenepodium quinoa willd.) ecotypes from there geogrophicd areas of Chile. Chilean Journal of Agricultural Research. 2012;**72**(2):175–181. DOI: 10.4067/S0718-58392012000200002

[30] James L.E.A. Chapter 1 Quinoa (Chenopodium quinoa Willd.): Composition, chemistry, nutritional, and functional properties. Advances in Food and Nutrition Research. 2009;**58**:1–31. DOI: 10.1016/S1043-4526(09)58001-1

[31] Huang AH, Turcot S, Graveline N, Pelletier M, Plourde H, Villeneuve S, Germain I. Nutritional and organoleptic characterization of two quinoa (Chenopodium quinoa) cultivars grown in Quebec, Canada. Foods. 2025;**14**:2394. DOI: 10.3390/foods14132394

[32] Repo-Carrasco R, Espinoza C, Jacobsen SE. Nutritional value and use of the Andean Crops Quinoa (Chenopodium quinoa) and Kañiwa (Chenopodium pallidicaule). Food Reviews International. 2003;**19**:179–189. DOI: 10.1081/FRI-120018884

[33] Chen J, Zhang Y, Guan X, Cao H, Li L, Yu J, Liu H. Characterization of Saponins from differently colored Quinoa cultivars and their In Vitro gastrointestinal digestion and fermentation properties. Journal of Agricultural and Food Chemistry. 2022;**70**:1810–1818. DOI: 10.1021/acs.jafc.1c06200

[34] Gao X, Yu B, Yu J, Mao X, Huang Z, Luo Y, Luo J, Zheng P, He J, Chen D. Effects of dietary starch structure on growth performance, serum glucose–insulin response, and intestinal health in weaned piglets. Animals. 2020;**10**(3):543. DOI: 10.3390/ani10030543

[35] Koh A, Vadder FD, Kovatcheva-Datchary P, Bäckhed F. From dietary fiber to host physiology: Short-chain fatty acids as key bacterial metabolites. Cell. 2016;**165**(6):1332–1345. DOI: 10.1016/j.cell.2016.05.041

[36] Mu H, Xue S, Sun Q, Shi J, Zhang D, Wang D, Wei J. Research progress of Quinoa Seeds (Chenopodium quinoa Wild.): Nutritional Components, technological treatment, and application research progress of quinoa seeds (Chenopodium quinoa Wild.): Nutritional components, technological treatment, and application. Foods. 2023;**12**:2087. DOI: 10.3390/foods12102087

[37] McCleary BV. Measurement of dietary Fiber: Which AOAC official method of analysis SM to use. Journal of

AOAC International. 2023. DOI: 10.1093/jaoacint/qsad051

[38] Lamothe LM, Srichuwong S, Reuhs BL, Hamaker BR. Quinoa (Chenopodium quinoa W.) and amaranth (Amaranthus caudatus L.) provide dietary fibres high in pectic substances and xyloglucans. Food Chemistry. 2015;**167**:490–496. DOI: 10.1016/j.foodchem.2014.07.022

[39] Liu J, Wang Z, Mai P, Hao Y, Wang Z, Wang J. Quinoa bran soluble dietary fiber ameliorates dextran sodium sulfate induced ulcerative colitis in BALB/c mice by maintaining intestinal barrier function and modulating gut microbiota. International Journal of Biological Macromolecules. 2022;**216**:75–85. DOI: 10.1016/j.ijbiomac.2022.06.194

[40] Priyanka M, Suneetha W, Maheswari K, Suneetha K, Kumari B. Glycemic index profiling of quinoa (Chenopodium quinoa Willd) variety. Journal of Pharmaceutical Research International. 2018;**21**(2):1–5

[41] Valencia-Chamorro SA. Quinoa: Overview. The world of food grains. In: Colin W, Harold C, Koushik S, Jon F, editors. Encyclopedia of Food Grains. Vol. **1**. Elsevier; Amsterdam: Academic Press; 2016. p. 341–348. https://api.pageplace.de/preview/DT0400.9780123947864_A26624452/preview-9780123947864_A26624452.pdf

[42] Tan M, Chang S, Liu J, Li H, Zhao Q. Physicochemical properties, antioxidant and antidiabetic activities of polysaccharides from quinoa (Chenopodium quinoa willd.) seeds. Molecules. 2020;**25**(17):3840. DOI: 10.3390/molecules25173840

[43] Tang Y, Li X, Chen PX, Zhang B, Hernandez M, Zhang H, Marcone MF, Liu R, Tsao R. Characterisation of fatty acid, carotenoid, tocopherol/tocotrienol compositions and antioxidant activities in seeds of three Chenopodium quinoa Willd. genotypes. Food Chemistry. 2015;**174**:502–508. DOI: 10.1016/j.foodchem.2014.11.040

[44] Zaaboul F, Liu Y. Vitamin E in foodstuff: Nutritional, analytical, and food technology aspects. Comprehensive Reviews in Food Science and Food Safety. 2022;**21**(2):964–998. DOI: 10.1111/1541-4337.12924

[45] USDA (U.S. Department of Agriculture). Quinoa nutritents [Internet]. 2019. Available from: https://fdc.nal.usda.gov/fdc-app.html#/food-details/168874/nutrients

[46] Pedrali D, Giupponi L, Pe˜na-Armada RD, Villanueva-su´arez MJ, Mateos-Aparicio I. The quinoa variety influences the nutritional and antioxidant profile rather than the geographic factors. Food Chemistry. 2023;**402**:133531. DOI: 10.1016/j.foodchem.2022.133531

[47] Angeli V, Miguel Silva P, Crispim Massuela D, Khan MW, Hamar A, Khajehei F, Graeff-Hönninger S, Piatti C. Quinoa (Chenopodium quinoa Willd.): An overview of the potentials of the "golden grain" and socio-economic and environmental aspects of its cultivation and marketization. Foods. 2020;**9**:216. DOI: 10.3390/foods9020216

[48] Graf BL, Silva PR, Rojo LE, Herrera JD, Baldeon ME, Raskin I. Innovations in health value and functional food development of quinoa (Chenopodium quinoa Willd.). Comprehensive Reviews in Food

Science and Food Safety. 2015;**14**:431–445. DOI: 10.1111/1541-4337.12135

[49] Tang Y, Tsao R. Phytochemicals in quinoa and amaranth grains and their antioxidant, anti-inflammatory, and potential health beneficial effects: A review. Molecular Nutrition and Food Research. 2017;**61**(7). DOI: 10.1002/mnfr.201600767

[50] Tang Y, Li X, Zhang B, Chen PX, Liu R, Tsao R. Characterisation of phenolics, betanins and antioxidant activities in seeds of three Chenopodium quinoa willd. Genotypes. Food Chemistry. 2015;**166**:380–388. DOI: 10.1016/j.foodchem.2014.06.018

[51] Gómez-Caravaca AM, Segura-Carretero A, Fernández-Gutiérrez A, Caboni MF. Simultaneous determination of phenolic compounds and saponins in quinoa (Chenopodium quinoa Willd) by a liquid chromatography–diode array detection–electrospray ionization–time-of-flight mass spectrometry methodology. Journal of Agricultural and Food Chemistry. 2011;**59**:10815–10825. DOI: 10.1021/jf202224j

[52] Wan MLY, Co VA, El-Nezami H. Dietary polyphenol impact on gut health and microbiota. Critical Reviews in Food Science and Nutrition. 2021;**61**(4):690–711. DOI: 10.1080/10408398.2020.1744512

[53] Espitia-Hernández P, Chávez González ML, Ascacio-Valdés JA, Dávila-Medina D. D, Flores-Naveda A, Silva T, Chacón XR, Sepúlveda L. Sorghum (Sorghum bicolor L.) as a potential source of bioactive substances and their biological properties. Critical Reviews in Food Science and Nutrition. 2022;**62**(8):2269–2280. DOI: 10.1080/10408398.2020.1852389

[54] Guo H, Wang S, Liu C, Xu H, Bao Y, Ren G, Yang X. Key phytochemicals contributing to the bitterness of quinoa. Food Chemistry. 2024;**449**:139262. DOI: 10.1016/j.foodchem.2024.139262

[55] Rafik S, Rahmani M, Rodriguez JP, Andam S, Ezzariai A, El Gharous M, Karboune S, Choukr-Allah R, Hirich A. How does mechanical pearling affect quinoa nutrients and saponin contents? Plants. 2021;**10**:1133. DOI: 10.3390/plants10061133

[56] Ruiz KB, Khakimov B, Engelsen SB, Bak S, Biondi S, Jacobsen SE. Quinoa seed coats as an expanding and sustainable source of bioactive compounds: An investigation of genotypic diversity in saponins profiles. Industrial Crops and Products. 2017;**104**:156–163. DOI: 10.1016/j.indcrop.2017.04.007

[57] El Hazzam K, Hafsa J, Sobeh M, Mhada M, Taourirte M, El Kacimi K, Yasri A. An insight into saponins from Quinoa (Chenopodium quinoa Willd): A review. Molecules. 2020;**25**(5):1059. DOI: 10.3390/molecules25051059

[58] Park JH, Lee YJ, Kim YH, Yoon KS. Antioxidant and antimicrobial activities of quinoa (Chenopodium quinoa Willd.) seeds cultivated in Korea. Preventive Nutrition and Food Science. 2017;**22**(3):195–202. DOI: 10.3746/pnf.2017.22.3.195

[59] Vilcacundo R, Hernández-Ledesma B. Nutritional and biological value of quinoa (Chenopodium quinoa Willd). Current Opinion in Food Science. 2017;**14**:1–6. DOI: 10.1016/j.cofs.2016.11.007

[60] Salehi B, Quispe C, Sharifi-Rad J, Cruz-Martins N, Nigam M, Mishra AP,

Konovalov DA, Orobinskaya V, Abu-Reidah IM, et al. Phytosterols: From preclinical evidence to potential clinical applications. Frontiers in Pharmacology. 2021;**11**:599959

[61] Graf B, Rojo LE, Delatorre-Herrera J, Poulev A, Calfio C, Raskin I. Phytoecdysteroids and flavonoid glycosides among Chilean and commercial sources of Chenopodium quinoa: Variation and correlation to physico-chemical characteristics. Journal of the Science of Food and Agriculture. 2015;**96**(2):633–643. DOI: 10.1002/jsfa.7134

[62] Sidorova YS, Shipelin VA, Petrov NA, Zorin SN, Maz VK. Anxiolytic and antioxidant effect of phytoecdysteroids and polyphenols from Chenopodium quinoa on an in vivo restraint stress model. Molecules. 2022;**27**:9003. DOI: 10.3390/molecules27249003

[63] Das N, Mishra SK, Bishayee A, Ali ES, Bishayee A. The phytochemical, biological, and medicinal attributes of phytoecdysteroids: An updated review. Acta Pharmaceutica Sinica B. 2021;**11**(7):1740–1766. DOI: 10.1016/j.apsb.2020.10.012

[64] Chaudhary N, Walia S, Kumar R. Functional composition, physiological effect and agronomy of future food quinoa (Chenopo-dium quinoa Willd.): A review. Journal of Food Composition and Analysis. 2023;**118**:105192. DOI: 10.1016/j.jfca.2023.105192

[65] Esatbeyoglu T, Wagner AE, Schini-Kerth VB, Rimbach G. Betanin—A food colorant with biological activity. Molecular Nutrition and Food Research. 2015;**59**:36–47. DOI: 10.1002/mnfr.201400484

[66] Gawlik-Dziki U, Świeca M, Sułkowski M, Dziki D, Baraniak B, Czyż J. Antioxidant and anticancer activities of Chenopodium quinoa leaves extracts-in vitro study. Food and Chemical Toxicology. 2013;**57**:154–160. DOI: 10.1016/j.fct.2013.03.023

[67] Francis G, Kerem Z, Makkar HPS, Becker K. The biological action of saponins in animal systems: A review. British Journal of Nutrition. 2002;**88**:587–605. DOI: 10.1079/BJN2002725

[68] Lin M, Han P, Li Y, Wang W, Lai D, Zhou L. Quinoa secondary metabolites and their biological activities or functions. Molecules. 2019;**24**(13):2512. DOI: 10.3390/molecules24132512

[69] Hernández-Ledesma B. Quinoa (Chenopodium quinoa Willd.) as source of bioactive compounds: A review. Bioactive Compounds in Health and Disease. 2019;**2**:27–47. DOI: 10.31989/bchd.v2i3.556

[70] Aluko RE, Monu E. Functional and bioactive properties of quinoa seed protein hydrolysates. Journal of Food Science. 2003;**68**(4):1254–1258. DOI: 10.1111/j.1365-2621.2003.tb09635.x

[71] Zamudio FV, M.r SC. Amaranth, quinoa and chia bioactive peptides: A comprehensive review on three an-cient grains and their potential role in management and prevention of Type 2 diabetes. Critical Reviews in Food Science and Nutrition. 2022;**62**(10):2707–2721. DOI: 10.1080/10408398.2020.1857683

[72] Schumacher A, Brandelli A, Macedo F, Pieta L, Klug T, Jong E. Chemical and sensory evaluation of dark chocolate with addition of quinoa

(Chenopodium quinoa willd.). Journal of Food Science & Technology. 2010;**47**(2):202–206. DOI: 10.1007/s13197-010-0029-x

[73] Tuo Y, Peng S, Li Y, Dang J, Feng Z, Ding L, Du S, Liu X, Wang L. Quinoa protein and its hydrolysate improve the fatigue resistance of mice: A potential mechanism to relieve oxidative stress and inflammation and improve energy metabolism. Journal of Nutritional Biochemistry. 2025;**139**:109863. DOI: 10.1016/j.jnutbio.2025.109863

[74] Pellegrini M, Lucas-Gonzales R, Ricci A, Fontecha J, Fernandez-Lopez J, Perez-Alvarez JA, Viuda-Martos M. Chemical, fatty acid, polyphenolic profile, techno-functional and antioxidant properties of flours obtained from quinoa (Chenopodium quinoa Willd) seeds. Industrial Crops and Products. 2018;**111**:38–46. DOI: 10.1016/j.indcrop.2017.10.006

[75] Lim JG, Park HM, Yoon KS. Analysis of saponin composition and comparison of the antioxidant activity of various parts of the quinoa plant (Chenopodium quinoa Willd.). Food Science and Nutrition. 2020;**8**(1):694–702. DOI: 10.1002/fsn3.1358

[76] Ilkay HO, Karabulut D, Kamaci Ozocak G, Mehmetbeyoglu E, Kaymak E, Kisioglu B, Cicek B, Akyol A. Quinoa (Chenopodium quinoa Willd.) supplemented cafeteria diet ameliorates glucose intolerance in rats. Food Science and Nutrition. 2023;**11**(11):6920–6930. DOI: 10.1002/fsn3.3603

[77] Omran NH, El-Bahy AA, Hosny HTA, Handoussa H. Quinoa and chia modulate AMPK/PPAR-γ signaling in high-fat diet- induced obesity rat model. Revista Brasileira de Farmacognosia. 2023;**33**(3):583–594. DOI: 10.1007/s43450-023-00388-5

[78] Yao Y, Yang XS, Shi ZX, Ren GX. Anti-inflammatory activity of saponins from quinoa (Chenopodium quinoa Willd.) seeds in lipopolysaccharide-stimulated RAW 264.7 macrophages cells. Journal of Food Science. 2014;**79**(5):H1018–23. DOI: 10.1111/1750-3841.12425

[79] Capraro J, De Benedetti S, Di Dio M, Bona E, Abate A, Corsetto PA, Scarafoni A. Characterization of chenopodin isoforms from quinoa seeds and assessment of their potential anti-inflammatory activity in caco-2 cells. Biomolecules. 2020;**10**(5):795. DOI: 10.3390/biom10050795

[80] Liu M, Zhu K, Yao Y, Chen Y, Li J. Antioxidant, anti-inflammatory and antitumor activities of phenolic compounds from white, red and black Chenopodium quinoa seed. Cereal Chemistry. 2020;**97**(3):703–713. DOI: 10.1002/cche.10286

[81] Li Y, Zhang Y, Tuo Y, You H, Li J, Wang L, Liu X, Ding L. Quinoa protein and its hydrolysate ameliorated DSS-induced colitis in mice by modulating intestinal microbiota and inhibiting inflammatory response. International Journal of Biological Macromolecules. 2023;**253**:127588. DOI: 10.1016/j.ijbiomac.2023.127588

[82] Zhong L, Lyu W, Lin Z, Lu J, Geng Y, Song L, Zhang H. Quinoa ameliorates hepatic steatosis, oxidative stress, inflamma-tion and regulates the gut microbiota in nonalcoholic fatty liver disease rats. Foods. 2023;**12**(9):1780. DOI: 10.3390/foods12091780

[83] Carvalho A, Marchini J, Navarro A. Metabolic parameters of postmenopausal

women after quinoa or corn flakes intake–a prospective and double-blind study. International Journal of Food Science and Nutrition. 2014;**65**(3):380–385. DOI: 10.3109/09637486.2013.866637

[84] Mahdavi AM, Javadivala Z, Khalili Y, Khalili M. Effects of Quinoa (Chenopodium quinoa) on inflammatory mediators: A systematic review of preclinical studies. Inflammopharmacology. 2023;**31** (1):241–251. DOI: 10.1007/s10787-022-01091-y

[85] Nonn L, Duong D, Peehl DM. Chemopreventive anti-inflammatory activities of curcumin and other phytochem-icals mediated by MAP kinase phosphatase-5 in prostate cells. Carcinogenesis. 2007;**28**:1188–1196. DOI: 10.1093/carcin/bgl241

[86] Pineli LLO, Botelho RBA, Zandonadi RP, Solorzano JL, Oliveira GT, Reis CEG, Teixeira DS. Low gly-cemic index and increased protein content in a novel quinoa milk. LWT—Food Science and Technology. 2015;**63**:1261–1267. DOI: 10.1016/j.lwt.2015.03.094

[87] Atkinson FS, Foster-Powell K, Brand-Miller JC. International tables of glycemic index and glycemic load values: 2008. Diabetes Care. 2008;**31**:2281–2283. DOI: 10.2337/dc08-1239

[88] Alvarez-Jubete L, Wijngaard H, Arendt EK, Gallagher E. Polyphenol composition and in vitro antioxidant activity of ama-ranth, quinoa buckwheat and wheat as affected by sprouting and baking. Food Chemistry. 2010;**119**(2):770–778. DOI: 10.1016/j.foodchem.2009.07.032

[89] Liu Y, Zhao C, Cai Y, Shue W. Effects of quinoa on blood glucose levels in diabetic mice. Acta Nutrimenta Sin. 2019;**41**:261–264. DOI: 10.26914/c.cnkihy.2019.001419

[90] Liu J, Wang Z, Wang Z, Hao Y, Wang Y, Yang Z, Li W, Wang J. Physicochemical and functional properties of soluble dietary fiber from different colored quinoa varieties (Chenopodium quinoa Willd). Journal of Cereal Science. 2020;**95**:103045. DOI: 10.1016/j.jcs.2020.103045

[91] Erfidan S, Dede S, Usta A, Yuksek V, Cetin S. The effect of quinoa (Chenopodium quinoa) on apoptotic, autophagic, anti-oxidant and inflammation markers in glucocorticoid-induced insulin resistance in rats. Molecular Biology Reports. 2022;**49**(7):6509–6516. DOI: 10.1007/s11033-022-07479-x

[92] Guo H, Hao Y, Richel A, Everaert N, Chen Y, Liu M, Yang X, Ren G. Antihypertensive effect of quinoa protein under simulated gastrointestinal digestion and peptide characterization. Journal of the Science of Food and Agriculture. 2020;**100**(15):5569–5576. DOI: 10.1002/jsfa.10609

[93] Hemalatha P, Bomzan DP, Sathyendra Rao BV, Sreerama YN. Distribution of phenolic antioxidants in whole and milled fractions of quinoa and their inhibitory effectson α-amylase and α-glucosidase activities. Food Chemistry. 2016;**199**:330–338. DOI: 10.1016/j.foodchem.2015.12.025

[94] Nongonierma AB, Maux SL, Dubrulle C, Barre C, Fitzgerald RJ. Quinoa (Chenopodium quinoa Willd.) protein hydrolysates with in vitro dipeptidyl peptidase IV (DPP-IV) inhibitory and antioxidant properties. Journal of Cereal Science. 2015;**65**:112–118. DOI: 10.1016/j.jcs.2015.07.004

[95] Li L, Hu X, Sun X, Wei X, Zhang H, Wang P. Quinoa alleviates the diabetic symptoms of db/db mice by upregulating insulin signaling and modulating gut microbiota composition. Journal of Functional Foods. 2025;**128**:106813. DOI: 10.1016/j.jff.2025.106813

[96] Diaz-Rizzolo DA, Acar-Denizli N, Kostov B, Roura E, Siso-Almirall A, Delicado P, Gomis R. Glycaemia fluctu-ations improvement in old-age prediabetic subjects consuming a quinoa-based diet: A pilot study. Nutrients. 2022;**14**(11):2331. DOI: 10.3390/nu14112331

[97] Ge Y, Shi Y, Wei C, Uthamapriya RA, Wu Y, Cao L.The effects of quinoa bran dietary fiber on glucose and lipid metabolism and hepatic transcriptome in obese rats. Journal of the Science of Food and Agriculture. 2024;**104**(5):2692–2703

[98] Zheng Y, Wang X, Zhuang Y, Li Y, Tian H, Shi P, et al Isolation of novel ace-inhibitory and antioxidant peptides from quinoa bran albumin assisted with an in-silico approach: Characterization, in vivo antihypertension, and molecular docking. Molecules. 2019;**24**:4562

[99] Marak NR, Das P, Das Purkayastha M, Baruah LD. Effect of quinoa (Chenopodium quinoa W.) flour supplementation in breads on the lipid profile and glycemic index: An in vivo study. Frontiers in Nutrition. 2024;**11**:1341539. DOI: 10.3389/fnut.2024.1341539

[100] Ajayi FF, Mudgil P, Jobe A, Antony P, Vijayan R, Gan C-Y, Maqsood S. Novel plant-protein (Quinoa) derived bioactive peptides with potential anti-hypercholesterolemic activities: Identi-fication, characterization and molecular docking of bioactive peptides. Foods. 2023;**12**(6):1327. DOI: 10.3390/foods12061327

[101] Woldemichael GM, Wink M. Identification and biological activities of triterpenoid saponins from Chenopodium quinoa. Journal of Agricultural and Food Chemistry. 2001;**49**:2327–2332. DOI: 10.1021/jf0013499

[102] Yu SH, Yu JM, Yoo HJ, Lee SJ, Kang DH, Cho YJ, et al Anti-proliferative effects of rutin on Oletf rat vascular smooth muscle cells stimulated by glucose variability. Yonsei Medical Journal. 2016;**57**:373–381

[103] Wolter A, Hager AS, Zannini E, Arendt EK. In vitro starch digestibility and predicted glycaemic indexes of buckwheat, oat, quinoa, sorghum, teff and commercial gluten-free bread. Journal of Cereal Science. 2013;**58**:431–436. DOI: 10.1016/jcs.2013.09.003

[104] Agarwal AR, Tripathi AD, Kumar T, Sharma KP, SK P. Nutritional and Functional New Perspectives and Potential Health Benefits of Quinoa and Chia Seeds. Antioxidants. 2023;**12**:1413. DOI: 10.3390/antiox12071413

[105] Deshmukh P, Unni S, Krishnappa G, Padmanabhan B. The Keap1–Nrf2 pathway: Promising therapeutic target to counteract ROS-mediated damage in cancers and neurodegenerative diseases. Biophysical Reviews. 2017;**9**:41–56. DOI: 10.1007/s12551-016-0244-4

[106] Imran M, Salehi B, Sharifi-Rad J, Gondal TA, Saeed F, Imran A, Shahbaz M, Fokou PVT, Arshad MU, Khan H, et al Kaempferol: A key emphasis to its anticancer potential. Molecules. 2019;**24**:2277

[107] El Makawy AI, Abdel-Aziem SH, Mohammed SE, Ibrahim FM, Abd EL-Kader HA, Sharaf HA, Youssef DA, Mabrouk DM. Exploration of tumor growth regression of quinoa and chia oil nanocapsules via the control of PIK3CA and MYC expression, anti-inflammation and cell proliferation inhibition, and their hepatorenal safety in rat breast cancer model. Bulletin of the National Research Centre. 2024;**48**(1):7. DOI: 10.1186/s42269-023-01161-3

[108] Fan X, Guo H, Teng C, Zhang B, Blecker C, Ren G. Anti-colon cancer activity of novel peptides isolated from in vitro digestion of quinoa protein in Caco-2 cells. Foods (Basel, Switzerland). 2022;**11**(2):194. DOI: 10.3390/foods11020194

[109] Ross AB, Zangger A, Guiraud SP. Cereal foods are the major source of betaine in the western diet-analysis of betaine and free choline in cereal foods and updated assessments of betaine intake. Food Chemistry. 2014;**145**:859–865. DOI: 10.1016/j.foodchem.2013.08.122

[110] Dong S, Yang X, Zhou L, Zhang F, Hou Z, Xue P. Antibacterial activity and mechanism of action saponins from Chenopodium quinoa willd. husks against foodborne pathogenic bacteria. Industrial Crops and Products. 2020;**149**:Article 112350. DOI: 10.1016/j.indcrop.2020.112350

[111] Sun X, Yang X, Xue P. Improved antibacterial effects of alkali-transformed saponin from quinoa husks against halitosis-related bacteria. BMC Complementary Medicine and Therapies. 2019;**19**:46. DOI: 10.1186/s12906-019-2455-2

[112] Stuardo M, Martín RS. Antifungal properties of quinoa (Chenopodium quinoa willd) alkali treated saponins against botrytis cinerea. Industrial Crops and Products. 2008;**27**(3):296–302. DOI: 10.1016/j.indcrop.2007.11.003

[113] Miranda M, Delatorre-Herrera J, Vega-Gálvez A, Jor-quera E, et al Antimicrobial potential and phytochemical content of six diverse sources of quinoa seeds (Chenopodium quinoa Willd.). Agricultural Sciences. 2014;**5**:1015–1024

[114] Ferdes M, Juhaimi FA, Ghafoor K. Inhibitory effect of some plant essential oils on growth of aspergillus Niger, aspergillus oryzae, mucor pusillus and fusarium oxysporum. South African Journal of Botany. 2017;**113**:457–460. DOI: 10.1016/.j.sajb.2017.09.020

[115] Yekta MM, Rezaei M, Nouri L, Azizi MH, Khaneghah AM. The antimicrobial and antioxidant properties of produced burgers with quinoa peptide- loaded nanoliposomes. Journal of Food Safety. 2019;**40**(2). DOI: 10.1111/jfs.12753

[116] Pagno CC, T.m.h. M, W. E, Benvenutti EV, Hertz PF, Matte CREA. Development of active biofilms of quinoa (Chenopodium quinoa w.) starch containing gold nanoparticles and evaluation of antimicrobial activity. Food Chemistry. 2015;**173**:755–762. DOI: 10.1016/j.foodchem.2014.10.068

[117] Mohamed DA, Fouda KA, Mohamed RS. In vitro anticancer activity of quinoa and safflower seeds and their preventive effects on non-alcoholic fatty liver. Pakistan Journal of Biological Sciences. 2019;**22**(8):383–392. DOI: 10.3923/pjbs.2019.383.392

[118] Wang M, Fu DY. Study on the mechanism of target organ damage in

hypertension basedonSirt1/Pgc1α/mitochondrialaxis. Chinese Journal of Traditional Chinese Medicine. 2020;**38**:139–144. DOI: 10.13193/j.issn.1673-7717.2020.04.033

[119] An T, Liu JX, Yang XY, Lv BH, Wu YX, Jiang GJ. Supplementation of quinoa regulates glycolipid metabolism and endoplasmic reticulum stress in the high-fat diet-induced female obese mice. Nutrition and Metabolism (London). 2021;**18**(1):95. DOI: 10.1186/s12986-021-00622-8

[120] Lin TA, Ke BJ, Cheng CS, Lee CL. Red quinoa bran extracts protect against carbon tetrachloride-induced liver injury and fibrosis in mice via activation of antioxidative enzyme systems and blocking TGF-1 pathway. Nutrients. 2019;**11**(2):395. DOI: 10.3390/nu11020395

[121] Ren G, Teng C, Fan X, Guo S, Zhao G, Zhang L, Liang Z, Qin P. Nutrient composition, functional activity and industrial applications of quinoa (Chenopodium quinoa Willd. Food Chemistry. 2023;**410**:135290. DOI: 10.1016/j.foodchem.2022.135290

[122] Zevallos VF, Herencia LI, Chang F, Donnelly S, Ellis HJ, Ciclitira PJ. Gastrointestinal effects of eating quinoa (Chenopodium quinoa Willd.) in celiac patients. American Journal of Gastroenterology. 2014;**109**(2):270–278. DOI: 10.1038/ajg.2013.431

[123] Demir M, Kılınç M. Quinoa: Nutritional and Anti-Nutritional Characteristics. Journal of Food Health Science. 2016;**2**(3):104–111. DOI: 10.3153/JFHS16011

[124] Elgeti D, Nordlohne SD, Föste M, Besl M, Linden MH, Heinz V, et al. Volume and texture improvement of gluten-free bread using quinoa white flour. Journal of Cereal Science. 2014;**59**(1):41–47. DOI: 10.1016/j.jcs.2013.10.010

[125] Li Y, Feng Z, Wu T, You H, Wang W, Liu X, Ding L. Quinoa peptides alleviate obesity in mice induced by a high-fat diet via regulating of the PPAR-α/γ signaling pathway and gut microbiota. Molecular Nutrition and Food Research. 2023;**67**(22): e2300258. DOI: 10.1002/mnfr.202300258

[126] Li W, Song Y, Cao YN, Zhang LL, Zhao G, Wu DT, Zou L. Total saponins from quinoa bran alleviate high-fat diet-induced obesity and systemic inflammation via regulation of gut microbiota in rats. Food Science and Nutrition. 2022;**10**(11):3876–3889. DOI: 10.1002/fsn3.2984

[127] Ruiz KB, Biondi S, Oses R, Acuña-Rodríguez IS, et al Quinoa biodiversity and sustainability for food security un-der climate change. A review. Agronomy for Sustainable Development. 2013;**34**:349–359

[128] Wang L, Wu S, Chen T, Xiong L, Wang F, Song H, Zhou J, Wei S, Ren B, Shen X. A quinoa peptide protects impaired mucus barriers in colitis mice by inhibiting NF-κB-TRPV1signaling and regulating the gut microbiota. Food Function. 2024;**15**(3):1223–1236. DOI: 10.1039/d3fo04905a

[129] Zhu F. Dietary fiber polysaccharides of amaranth, buckwheat and quinoa grains: A review of chemical structure, biological functions and food uses. Carbohydrate Polymers. 2020;**248**:116819. DOI: 10.1016/j.carbpol.2020.116819

[130] Zang Y, Ge Y, Cao Y, Tang H. Anti-diabetic effect of red quinoa

polysaccharide on type 2 diabetic mellitus mice induced by streptozotocin and high-fat diet. Frontiers in Microbiology. 2024;**15**:1308866. DOI: 1 0.3389/fmicb.2024.1308866

[131] Han Y, Chi J, Zhang M, Zhang R, Fan S, Dong L, Huang F, Liu L. Changes in saponins, phenolics and antioxidant activity of quinoa (Chenopodium quinoa willd) during milling process. LWT. 2019;**114**:108381. DOI: 10.1016/ jlwt.2019.108381

[132] Lorusso A, Verni M, Montemurro M, Coda R, Gobbetti M, C. G R. Use of fermented quinoa flour for pasta mak-ing and evaluation of the technological and nutritional features. LWT. 2017;**78**:215–221. DOI: 10.1016/j. lwt.2016.12.046

[133] Sánchez-García J, Asensio-Grau A, García-Hernández J, Heredia A, Andrés A. Nutritional and antioxidant changes in lentils and quinoa through fungal solid-state fermentation with Pleurotus ostreatus. Bioresources and Bioprocessing. 2022;**9**(1):51. DOI: 10.11 86/.s40643-022-00542-2

[134] Carrizo SL, LeBlanc AM, LeBlanc JG, Rollan GC. Quinoa pasta fermented with lactic acid bacteria prevents nutritional deficiencies in mice. Food Research International. 2020;**127**:108735. DOI: 10.1016/j. foodres.2019.108735

[135] Chu R, Uaila E, Ismail T, Lazarte CE. Effect of short-term lactic fermentation on polyphenol profile and antioxidant capacity in white and red quinoa varieties. Foods. 2024;**13** (15):2413. DOI: 10.3390/foods13152413

[136] Connolly DA. Impact of quinoa and food processing methods on the human gut microbiome through in vitro fermentation. ProQuest. 2023. https://rex.libraries.wsu.edu/ esploro/outputs/graduate/Impact-of-Quinoa-and-Food-Processing/999010 31139101842

[137] Mhada M, Metougui ML, Hazzam KE, Kacimi KE, Yasri A. Variations of saponins, minerals and total phenolic compounds due to processing and cooking of quinoa (Chenopodium quinoa Willd.) seeds. Foods. 2020;**9**(5):660. DOI: 10.3390/ foods9050660

[138] Han Y, Chi J, Zhang M, Zhang R, Fan S, Dong L, Huang F, Liu L. Changes in saponins, phenolics and antioxidant activity of quinoa (Chenopodium quinoa willd) during milling process. LWT. 2019;**114**:108381. DOI: 10.1016/j. lwt.2019.108381

[139] Kowalski RJ, Medina-Meza IG, Thapa BB, Murphy KM, Ganjyal GM. Extrusion processing characteristics of quinoa (Chenopodium quinoa Willd.) var. Cherry Vanilla. Journal of Cereal Science. 2016;**70**:91–98. DOI: 10.1016/j. jcs.2016.05.024

[140] Jan R, Saxena DC, Singh S. Effect of extrusion variables on antioxidant activity, total phenolic content and dietary fibre con-tent of gluten-free extrudate from germinated Chenopodium (Chenopodium album) flour. International Journal of Food Science & Technology. 2017;**52** (12):2623–2630. DOI: 10.1111/ijfs.13549

[141] Bhinder S, Kumari S, Singh B, Kaur A, Singh N. Impact of germination on phenolic composition, antioxidant properties, antinutritional factors, mineral content and Maillard reaction prod-ucts of malted quinoa flour. Food Chemistry. 2021;**346**:128915. DOI: 10.1 016/j.foodchem.2020.128915

[142] El-Deeb AM, Hassan NSY, Hassanein AM. Preparation and properties of flavored fermented beverage based on partial or complete replacement of milk with Quinoa Seeds Water Extract (QSWE). International Journal of Dairy Science. 2014;**9** (4):96–105. DOI: 10.3923/ijds.2014.96.105

[143] Abugoch LE, Tapia C, Villamán MC, Yazdani-Pedram M, Díaz-Dosque M. Characterization of quinoa protein–chitosan blend edible films. Food Hydrocolloids. 2011;**25**:879–886. DOI: 10.1016/j.foodhyd.2010.08.008

Section 4

Climate-Resilient Breeding and Crop Wild Relatives

Chapter 6

Use of Crop Wild Relatives in Wheat for Climate Change and Food Security

Anand Kumar, Laxmidas Verma, Ravindra Kumar and Sagar

Abstract

Global food security relies on the ongoing enhancement of a limited number of key food and feed crops. The current food supply is significantly dependent on technology. To feed over 9 billion people by 2050, we must double high-quality food production while ensuring environmental and social sustainability. Current crop improvement practices are insufficient to meet future demands or address the impact of climate change. Integrating food security and climate change strategies presents a promising approach to enhance crop resilience against biotic and abiotic stresses. Wheat is a major staple crop worldwide; wheat yield gains were achieved through increased agricultural inputs and the expansion of arable land. However, future improvements hinge on genetic enhancements, which have declined due to reduced genetic diversity in wheat breeding. This lack of variability poses a challenge for further advancements in technology. Moreover, using both wild and cultivated relatives of wheat offers new opportunities for genetic improvement in wheat production. Innovations in genetic enhancements for wheat breeding are vital for addressing climate change challenges and ensuring food security. Identifying and utilizing these advancements will revolutionize wheat cultivation, enabling precise improvements in crop resilience and productivity. There are several conventional techniques used to improve wheat productivity and resistance to biotic and abiotic stress. However, these techniques are more expensive and time-consuming. Here, in this book chapter, we describe the use of crop wild relatives and advancements in technology for global food security.

Keywords: wheat, global food security, crop wild relatives (CWR), biotic and abiotic stress and advancement of technology, climate change

1. Introduction

Mankind must confront the truth of global warming. Food security is the most essential need for all life on Earth, and it is significantly affected by climate change. This makes it a pressing issue that requires urgent action [1]. Climate change, fluctuations in weather during the growing season, adaptation strategies that help crops realize their genetic potential in response to a changing climate, and the creation of cultivars better suited to varying environmental conditions are all factors that will influence future food security [2]. Since our capacity to foretell future climates is uncertain, these problems may seem insurmountable. Despite these obstacles, we can learn more about the interplay between soil, plants, and the atmosphere, and how to use this information to further our aim of universally improved food security [3].

Crop wild relatives (CWR) are plant species that are closely related to field crops and their ancestors. These species can contribute beneficial traits to crop improvement, such as resistance to biotic and abiotic stresses, and they can enrich the gene pool, which ultimately leads to enhanced plant yield [4]. This helps humanity in its relentless search for food production to meet the ever-growing demands of a burgeoning global population. Actually, CWRs are well-known for their immense potential to maintain and improve global food security, which would greatly benefit humanity [5]. This underscores the vital importance of identifying, describing, and preserving these elements in crop breeding efforts. The significance of this work is further amplified in light of the recent challenges posed by global climate change [6].

Given the severe consequences, particularly the concerning environmental risks, of the agricultural practices that heavily relied on soil extraction throughout the 1960s, sometimes referred to as the "Green Revolution," the task becomes even more important [7].

Climate change poses a serious risk to humankind's ability to grow food. Greenhouse gas emissions from human activities are rapidly altering the environment and accelerating the process of global warming. Soil water deficiencies will worsen, and organic matter decomposition will speed up as a result of this in the future [8]. Soil erosion will be a major issue unless we put conservation measures in place to address the rising frequency of heavy rainfall, runoff, and flooding in the future [9]. Conservative agricultural practices, such as reduced or no tillage, crop residue retention, and diverse cropping systems, can help reduce atmospheric CO_2 levels, according to a new mitigation strategy that aims to transfer the gas into soil organic carbon. The reach of these policies is limited, nevertheless, due to the many interregional and international constraints that prevent them from being implemented [10].

Putting the two trends of rising populations and average global temperatures side by side paints a bleak picture. According to current predictions, the global temperature is expected to increase by 1.8–4°C, and the global population could reach 10 billion by 2100, a 142.9% increase from the present 7 billion, or around a 1.8% per year increase [5]. Two types of "greenhouse effects" exist: one, which is "natural," makes certain high-latitude and high-altitude regions habitable for life, and the other, which is far more common and much more harmful to the environment, is caused by industrial gases like carbon dioxide and agricultural practices like the overuse and unchecked use of nitrogen fertilizers such as urea. These practices lead to huge concentrations of nitrous oxide in the air and are killing off plants and humans alike [11]. There has been a 2000% increase in the atmospheric concentration of carbon dioxide (CO_2) from its pre-industrial value of 20 ppm to 400 ppm, with a significant portion of this increase attributable to emissions from automobiles [12], an absolutely astounding growth. The most worrisome aspect, however, is the generation of nitrous oxide (N_2O), also called "laughing gas" (dinitrogen monoxide), which is a key contributor to the depletion of stratospheric ozone and, by extension, to climate change. As a "greenhouse gas," nitrous oxide has the third-greatest impact on global warming, after carbon dioxide and methane (CH_4) [13]. Nitrous oxide (N_2O) is 310 times more effective at trapping heat than carbon dioxide (CO_2). It typically remains in the atmosphere for about 120 years. The excessive use of nitrogen-based fertilizers, like urea, is responsible for a significant portion of agricultural emissions, contributing to 30% of the N_2O in the atmosphere. During the 1960s, coinciding with the Green Revolution, there was a notable rise in atmospheric nitrogen oxides (NO_x) [14]. When nitrogen fertilizers, such as urea, are used in excess with the goal of increasing food production, soil bacteria convert the nitrogen in the fertilizer to nitrous oxide at a greater rate than usual [15]. All the environmental consequences of global warming, such as reduced precipitation, soil deterioration, groundwater depletion, etc., have their roots in this. This ecological catastrophe is on full display in the Indian state of Punjab, sometimes referred to as the "cradle" of the Green Revolution, where thousands of acres of arable land have been eradicated [16].

Preserving CWR and extracting potentially useful genes to boost agricultural output is crucial, especially given the growing concerns about the devastating impacts of climate change on biodiversity and food security, alongside the world's rapidly increasing population [17]. To adapt domestic crops to climate change, CWRs are an important tool to overcome the restrictions of genetic variety. A lack of knowledge about their diversity, present and future value, and practical ways to conserve them is preventing their integration into conservation and encouraging greater systematic exploitation [18].

Environmental shifts and the movement of birds: Israeli specialists have found that, due to climate change, Israel is now serving as a permanent wintering ground for some 500 million migrating birds. These birds used to make brief stops before heading to the warmer plains of Africa. However, instead of traveling to Africa, where food is becoming harder to find due to expanding deserts and frequent droughts, they are now choosing to stay longer in milder areas [19]. Ornithologist Shay Agmon, who is in charge of the wetlands sector of Agamon Hula in northern Israel, stated that "many more birds and a greater number of species can no longer cross the desert," meaning that Israel has become more than simply a brief stopover for the birds in recent decades [20]. Over time, their duration of stay will increase, and the overall pattern of migration will undergo a transformation. Migrating birds may be fascinating to observers of birds and visitors, but they pose a threat to farmers due to their voracious appetite for crops. In the Syrian-African Rift Valley, in the Hula reserve, workers have made life more comfortable for the birds by feeding them in the marshes and luring them away from the fields [21]. A considerably wider desert is more difficult for the birds to fly across, and they just cannot do it. According to Agmon, fuel is running low, and there aren't enough "gas stations" along the route; thus, Israel has become a major hub for them. The migratory patterns of birds can teach us something about how to deal with the problems of climate change, which threatens the entire global food chain. Worldwide, regionally, and even locally, climatic change, stress, and variability [22] are the primary factors that determine biodiversity and evolutionary change [23]. Perhaps no other environmental threat that the world confronts today compares to the severity of global warming. An assessment of this issue is given in a recent report by an intergovernmental panel on climate change [24]. Climate change has been studied by [25] in relation to agricultural and wildlife populations and by [26] in terms of bananas, barley, beans, cassava, chickpeas, pigeonpeas, potatoes, faba beans, groundnuts, lentils, maize, millet, rice, sorghum, soybeans, wheat, yam, forages, fisheries, and aquaculture. From a scientific, political, and economic perspective, the most controversial subject today is global warming and its causes, which were briefly touched upon in the introductory chapter. There is still noticeable disagreement among some global powers, especially the United States, but everyone agrees that if we don't do something about climate change, it will have a devastating impact on farming, just like the very soil-exploitative Green Revolution had on the poor [27].

Domestication of plants and animals is the first step in the evolutionary chain that, after millennia, culminated in modern human society. Darwin postulated in 1905 that domestication is a massive evolutionary experiment that produces new species through adaptation and speciation. Like speciation in nature, it has mostly been done by humans in the last 10,000 years [28].

Adaptive syndromes that are suitable for human ecology are the result. Humans were permanently settled into cities, developed sophisticated cultural practices, and saw a massive increase in population as a result of domestication and the shift from pre-agricultural to agricultural economies. Domesticated adaptation syndromes in wheat involve higher yields and spikes that remain intact. Due to the process of domestication, these cultivars have become culturally and economically reliant on humans, thriving only in specific agricultural environments shaped by human activity [29]. One of the most detrimental outcomes of domestication, in hindsight, was the complete failure of the so-called Green Revolution and the environmental risks it introduced due to the overuse of artificial fertilizers on already fertile soils [30]. The sad commentary of this collapse is seen in the countries of South Asia, especially India and, more specifically, in the Indian state of Punjab.

2. A brief history of CWR in wheat

Crop wild relatives (CWRs) are wild plant species closely related to cultivated crops. They serve as vital reservoirs of genetic diversity that have significantly contributed to agricultural advancements. In the case of wheat (Triticum spp.), CWR have been crucial in boosting yields, improving disease resistance, and adapting to various environmental conditions [31]. From the early domestication of wheat in the Fertile Crescent to the cutting-edge genomics of the 21st century, CWR have been a cornerstone of wheat improvement. As global challenges mount, the importance of these genetic resources continues to grow, underscoring the need for their conservation and sustainable use [32].

2.1 Early domestication and role of wild relatives

The history of wheat begins around 10,000 years ago in the Fertile Crescent, where early humans first domesticated *Triticum dicoccum* (emmer wheat) and *Triticum monococcum* (einkorn wheat). These species evolved from wild relatives like *Triticum urartu* and *Aegilops tauschii*. The hybridization of these wild ancestors gave rise to hexaploid wheat (*Triticum aestivum*), which is the basis of modern bread wheat [33].

During this domestication process, early farmers unintentionally selected for beneficial traits. However, domestication also resulted in a genetic bottleneck, where much of the genetic diversity present in wild species was lost. This made cultivated wheat less resilient to environmental changes, diseases, and pests [34].

2.2 19th and early 20th century: Rediscovery of CWR

The systematic use of CWR in wheat breeding began in the late 19th and early 20th centuries. Botanists and plant breeders recognized the potential of wild relatives as sources of resistance to diseases such as rusts (caused by *Puccinia* spp.) [35].

- **Nikolai Vavilov's contribution**: The Russian geneticist Nikolai Vavilov was among the first to identify the importance of CWR. In the 1920s and 1930s, Vavilov's expeditions across the world helped collect a vast array of wild wheat species and landraces. He proposed the "Centers of Origin" theory, emphasizing the significance of wild relatives in the genetic improvement of crops [36].

- ***Aegilops tauschii* in bread wheat**: The discovery of *Aegilops tauschii* as the donor of the D-genome in hexaploid wheat highlighted its importance in breeding. This wild relative provided genetic material crucial for traits like cold tolerance and resistance to certain pathogens [37].

2.3 Mid-20th century: Green revolution and beyond

The Green Revolution of the 1940s–1960s, led by scientists like Norman Borlaug, brought about a massive increase in wheat production. While much of this success was due to semi-dwarf varieties and improved agronomic practices, the role of CWR in wheat improvement became increasingly evident [38].

- **Disease resistance**: During this period, wild relatives were intensively used to introduce resistance genes. For example, genes from *Aegilops spp.* and *Secale cereale* (rye, another distant relative) were incorporated into wheat to combat rust diseases [39].

- **Synthetic wheat development**: The concept of synthetic wheats emerged in the mid-20th century. By crossing diploid and tetraploid species with *Aegilops tauschii*, breeders recreated the hybridization events that originally gave rise to hexaploid wheat. These synthetic wheats introduced novel genetic diversity into the wheat gene pool, enhancing traits like drought tolerance and pest resistance [40].

2.4 Late 20th century: Advances in molecular biology

The advent of molecular biology and genomics in the 1980s and 1990s revolutionized the use of CWR in wheat breeding. Scientists were now able to precisely identify and transfer specific genes from wild relatives into cultivated varieties [41].

- **Marker-assisted selection (MAS)**: This technique enables breeders to track the incorporation of beneficial genes from CWR into wheat, reducing the time and effort required for traditional breeding.

- **Translocations and introgressions**: Segments of chromosomes from wild relatives have been successfully introduced into wheat. For example, the wheat-rye translocation (1BL/1RS) has brought in resistance to various diseases and enhanced yield potential [42].

- **Notable genes**:

    - The *Lr* (Leaf rust) and *Sr* (Stem rust) gene series from CWR, such as *Sr31*, provide long-lasting resistance against devastating rust outbreaks.

    - The *Gpc-B1* gene, derived from wild emmer wheat, enhances grain protein content.

2.5 21st century: Climate change and renewed interest in CWR

The increasing challenges of climate change, a growing population, and the need for sustainable agriculture have reignited interest in CWR. Modern wheat breeding programs focus on traits like heat tolerance, water-use efficiency, and resistance to emerging pests and pathogens.

- **Global conservation efforts**: Organizations like the Global Crop Diversity Trust (Crop Trust) and initiatives such as the Svalbard Global Seed Vault are safeguarding CWR for future use. The International Maize and Wheat Improvement Center (CIMMYT) and other research institutions maintain vast germplasm collections, including CWR.

- **Genome editing and genomics**: The application of CRISPR-Cas9 and other genome-editing tools has made the precise transfer of genes from CWR into wheat a reality. Researchers are now able to directly target traits of interest without disrupting other aspects of the plant's genome.

- **New frontiers**: Efforts are underway to domesticate CWR themselves, effectively turning wild species into viable crops. Additionally, advanced genomic tools have revealed hidden diversity within wild species, opening new possibilities for wheat improvement.

2.6 Challenges in using CWR

Despite their potential, integrating CWR into wheat breeding is not without its challenges.

- **Genetic linkage drag**: Transferring desirable traits often brings along unwanted characteristics.

- **Crossing barriers**: Some CWR are difficult to hybridize with cultivated wheat due to ploidy differences or genetic incompatibilities.

- **Habitat loss**: The natural habitats of many CWRs are under threat from human activities, reducing the availability of these valuable genetic resources.

3. Use of advance technology for biotic and abiotic resistance

The genetic variability of wheat is increasingly limited due to reliance on conventional breeding methods involving landraces and common varieties. This narrowing gene pool poses challenges in addressing emerging biotic and abiotic stresses driven by climate change, such as salinity, waterlogging, drought, heat, and pests. Broadening the genetic base is critical for enhancing productivity and yield stability. Wild wheat progenitors harbor valuable genetic variations that can strengthen breeding pools. Utilizing these resources through conventional breeding and wide-crossing techniques, including interspecific, intraspecific, and intergeneric hybridization, is essential for improving wheat resilience and adaptability [43].

3.1 Biotic stress

Biotic stress in plants arises from living organisms such as fungi, viruses, insects, nematodes, arachnids, and weeds. Unlike abiotic stresses caused by environmental factors, biotic stresses directly affect plant growth and development by depriving them of essential nutrients, which can lead to reduced vigor and even death. In agriculture, these stresses are major contributors to pre- and post-harvest losses, significantly impacting crop productivity and food security [44].

The rising global population and increasing abiotic stresses, exacerbated by climate change, are compounded by significant biotic stresses that threaten wheat production worldwide. Diseases like rusts and mildew, along with pests such as aphids, drastically reduce the yield potential of elite wheat cultivars. The complex nature of plant-parasite interactions is pivotal in determining resistance expression. Biotic stress in plants is caused by living organisms such as fungi, viruses, insects, nematodes, arachnids, and weeds. Unlike abiotic stresses (e.g., heat and drought) caused by environmental factors, biotic stress agents directly impact host growth and development by depriving plants of essential nutrients, leading to reduced vigor and, in severe cases, plant death. In agriculture, biotic stresses significantly contribute to both pre- and post-harvest losses, posing a major challenge to crop productivity and food security [45].

3.1.1 Leaf and stem rust

Wheat leaf rust, caused by *Puccinia triticina* Eriks., wheat stem rust by *Puccinia graminis* f. sp. *tritici*, and wheat stripe rust by *Puccinia striiformis* f. sp. *tritici*, are among the most significant diseases affecting wheat crops. Among these, leaf rust occurs more frequently and in a wider range of regions globally compared to stem rust and stripe rust. Yield losses due to *P. triticina* have a considerable economic impact; infections typically result in a reduced number of kernels per head and decreased kernel weight, leading to yield losses that can reach up to 40% in highly susceptible cultivars [46]. Yield losses caused by stem rust pathogens during the mid-20th century reached 20–30% in Eastern and Central Europe, as well as in many other countries, including Australia, China, and India [47, 48].

3.2 Abiotic stress (e.g., drought tolerance, heat tolerance, salinity, and heavy metals tolerance)

In recent years, global warming and climate change have significantly impacted agricultural productivity in tropical and subtropical regions, leading to the emergence of new abiotic stresses. Among these, heat, drought, moisture, and salt stress are the most prevalent. This section provides insights into the effects of various abiotic stresses on the growth and physiology of wheat [49].

3.2.1 High temperature

Heat is one of the most common abiotic stresses affecting wheat. This section explores its impact on wheat growth and physiology. Wheat is particularly vulnerable to heat stress during specific physiological growth stages [50]. Heat stress, caused by elevated temperatures, is defined as an increase in air temperature beyond a critical threshold for a specific duration, leading to injury or irreversible damage in crop plants. A temperature rise of 1°C can result in a yield loss of approximately 4.1% to 6%. Studies have shown that heat stress induces pollen sterility, reduces carbon dioxide assimilation, and enhances photorespiration in wheat. Additionally, high temperatures impair the photosynthetic process, negatively affecting wheat growth and yield [51].

3.2.2 Drought stress

Drought stress, characterized by water deficiency, triggers significant morphological, biochemical, physiological, and molecular changes, making plant growth and crop production challenging. It negatively affects wheat at all developmental stages, with the severity of its impact depending on stage-specific stress and local climate conditions [52]. More than half of the world's wheat-growing regions experience drought regularly, leading to production losses of up to 10%. Additionally, global warming is expected to exacerbate drought frequency and intensity due to declining precipitation and increased evaporation.

3.2.3 Salinity stress

Salinity is one of the most detrimental abiotic stresses, significantly reducing agricultural productivity. It severely impacts plant morphology, physiology, and biochemistry, affecting processes such as photosynthesis, nutrient uptake, and yield. Studies have shown that wheat cultivars exhibit varying degrees of salt tolerance, with some being more resistant, while others are highly susceptible. High salinity levels hinder germination and increase Na^+ and Cl^- ion concentrations, disrupting the normal metabolic processes of wheat plants.

Salt stress affects plants through osmotic imbalance, ion toxicity, and nutritional disorders. In response to salinity stress, researchers assess seedling growth, seed reserve utilization, and biomass weight. Studies on different wheat cultivars show that salinity reduces relative water content, chlorophyll and carotenoid levels, biomass, and grain yield, likely due to increased oxidative stress. Under 150 mM NaCl stress, key physiological parameters, such as net photosynthetic rate, transpiration rate, stomatal conductance, and sub-stomatal carbon dioxide concentration decline significantly [53].

When plants are exposed to environmental stresses such as drought, salinity, and heat, they generate a large number of reactive oxygen species (ROS), which are biosynthesized in the chloroplast, mitochondria, and/or peroxisomes of plant cells. Reactive oxygen species (ROS) are highly reactive, and their excessive production can disrupt normal plant metabolism. High ROS concentrations can damage plant cells by promoting lipid peroxidation, protein degradation, and DNA denaturation, leading to oxidative stress. This, in turn, can impair enzymatic activity, disrupt cellular structures, and even result in cell death. Another stress metabolite, methylglyoxal (MG), is produced in plant chloroplasts, mitochondria, and the cytosol under abiotic stress conditions. MG is toxic to plant cells, contributing to oxidative stress and disrupting cellular functions. Its impact depends on its concentration and the severity of the stress. Among abiotic stresses, drought stress disrupts the balance between antioxidant defenses and drought-induced ROS production. Malondialdehyde (MDA) production is considered a marker of ROS generation and oxidative damage. Elevated levels of H_2O_2 and MDA in wheat cells disrupt adaptive mechanisms under drought-induced oxidative stress. Similarly, salinity stress is a major environmental constraint that inhibits plant growth and physiological functions by triggering excessive ROS production. Increased MDA accumulation in wheat cultivars under salinity stress leads to oxidative damage in parenchymal cells [54].

3.2.4 Advances in molecular breeding for wheat improvement

While traditional breeding has played a significant role in enhancing wheat yields, modern molecular breeding techniques have transformed the creation of superior wheat varieties. Some important molecular methods include:

1. Genomic selection (GS): Genomic selection employs high-throughput DNA sequencing and statistical models to estimate the breeding value of wheat plants based on their genetic composition. This method enables breeders to identify the most promising candidates for breeding, thereby shortening the time needed for variety development [55].

2. CRISPR/Cas9 genome editing: CRISPR/Cas9 has become a powerful tool for making precise genetic changes in wheat. This technology allows researchers to modify genes that influence yield, stress tolerance, and disease resistance, thus speeding up the breeding process [56].

3. Transgenic approaches: Genetic engineering facilitates the incorporation of advantageous genes from other organisms into wheat. For instance, genes that provide resistance to fungal pathogens or enhance nitrogen-use efficiency have been successfully integrated into wheat using transgenic methods [57].

3.3 Current status of CWR usage for abiotic stress tolerance in wheat

The primary gene pool of wheat consists of its direct progenitors, which can be easily crossed with cultivated wheat to produce fertile offspring. These include *Triticum urartu*, *Aegilops speltoides*, and *Aegilops tauschii*. The secondary gene pool comprises wild relatives, such as *Triticum turgidum* ssp. *dicoccoides* and *Triticum timopheevii*, which can hybridize with wheat, though the resulting progeny are often infertile (Pour-Aboughadareh et al., 2021). The tertiary gene pool includes more distantly related species, such as *Thinopyrum ponticum* and *Thinopyrum distichum*, which typically require advanced breeding techniques, such as colchicine treatment or embryo rescue, to produce viable progeny [58].

Crop wild relatives (CWR) are valuable plant genetic resources (PGR) widely utilized by breeders. Sharing a close evolutionary relationship with their domesticated counterparts, CWR have retained higher genetic diversity due to their avoidance of domestication [59]. Ex situ conservation of CWR plays a crucial role in preserving this diversity and serves as a vital resource for introducing tolerance to abiotic stress factors in crop improvement programs. Climate change intensifies environmental (abiotic) stress, significantly impacting crop production. Rising global temperatures alter precipitation patterns, leading to increased rainfall in some regions and drought in others [60]. Additionally, higher temperatures accelerate evapotranspiration, further exacerbating water stress and affecting crop growth and yield. *Thinopyrum distichum* (Thunb.) Löve, commonly known as sea wheat, is a halophytic wild relative of wheat. Native to the coastal dunes of South Africa, this species thrives in saline environments, making it a valuable genetic resource for improving salt tolerance in wheat breeding. Extensive research has identified three key mechanisms that plants employ to tolerate high salt levels: osmotic adjustment, ion homeostasis, and ROS homeostasis. These mechanisms collectively help plants maintain cellular function and sustain growth under saline conditions [61].

Although, *Thinopyrum distichum* is known for its inherent tolerance to saline environments, the specific mechanisms it employs to manage high salt stress remain unidentified. Further research is needed to determine the physiological and molecular pathways underlying its salt tolerance [62]. Advances in molecular breeding techniques, including genomics, MAS, transgenic approaches, and gene editing, have greatly accelerated the development of salt-tolerant wheat varieties. Genome sequencing and the identification of quantitative trait loci (QTLs) associated with salinity tolerance enable more precise selection, while genetic engineering and CRISPR/Cas9 technology facilitate targeted gene modifications to enhance salt resistance [63]. Additionally, introgression breeding incorporating beneficial traits from wild wheat relatives further strengthens wheat's resilience to saline conditions. A comprehensive understanding of the genetic and phenotypic diversity within local Thinopyrum distichum populations can enhance its potential for integration into wheat breeding programs, facilitating the development of salt-tolerant wheat varieties [64]. Previous research at Stellenbosch University's Plant Breeding Laboratory on Thinopyrum distichum has involved hybridization efforts, exploring its potential for improving wheat through breeding. The salt tolerance potential of perennial grasses has been studied since the 1960s, when Agropyron elongatum was identified as the most salt-tolerant CWR species at the time [61].

Conservation and utilization of CWRs

1. Conservation strategies: To ensure the ongoing availability of CWRs, effective conservation efforts are essential. These efforts include:

   1.a. In situ conservation: Safeguarding CWRs in their natural environments by establishing protected areas and biodiversity reserves.

   1.b. Ex situ conservation: Storing CWRs in gene banks and seed vaults, such as the Svalbard Global Seed Vault and the International Center for Agricultural Research in the Dry Areas (ICARDA).

   1.c. Community-based conservation: Involving local farmers and indigenous communities in the conservation of CWRs [65].

2. Challenges in CWR conservation and utilization

Despite their importance, several challenges impede the conservation and utilization of CWRs:

Habitat destruction: Urban development, deforestation, and changes in land use pose threats to the natural habitats of CWRs.

Limited research and funding: Many CWRs are under-researched due to a lack of funding for breeding programs and conservation efforts.

Crossbreeding barriers: Some CWRs are genetically distant from cultivated crops, necessitating advanced breeding techniques like MAS and gene editing.

Case studies: Success stories of CWR utilization

The introduction of wild wheat genes into commercial varieties: Researchers have successfully integrated genes from wild wheat species (*Aegilops tauschii*) to improve disease resistance and stress tolerance in modern wheat varieties. These initiatives have led to increased wheat yields in areas prone to drought [66].

4. Food and nutrition security challenge

Food security, as defined by the UN's FAO, exists when all people, at all times, have physical, social, and economic access to sufficient, safe, and nutritious food that meets their dietary needs and preferences for an active, healthy life. This means food must not only be available but also safe to eat and nutritionally adequate. In many Western countries, where food abundance has been the norm for decades, food safety has become a greater concern than availability [67]. Plant breeders must recognize this shift, as consumer safety is paramount. Additionally, the food industry is highly sensitive to reputational risks, and companies will quickly adapt – whether by changing suppliers or raw materials – to mitigate the chances of negative publicity. The formation of the processing contaminant acrylamide during high-temperature cooking and processing of foods made from potatoes, cereals, and other crops is a significant food safety concern. The food industry is highly sensitive to such issues and seeks to minimize risks, making any increase in acrylamide levels particularly unwelcome [68].

The relationship between food security and nutrition security is complex, as evidenced by the link between food insecurity and malnutrition outcomes such as overweight and obesity. Data indicate that obesity is more prevalent in food-insecure populations for several reasons. Limited access to healthy foods can contribute to weight gain through multiple pathways. For instance, low-income families often prioritize cost-effective, energy-dense foods over more expensive, nutrient-rich options to maximize their limited resources [69].

Evidence suggests that food-insecure households often prioritize food quantity over quality or variety to prevent absolute hunger. This can lead to nutrition insecurity despite an abundance of calories, particularly if households lack access to a consistent and adequate diet. Additionally, when food is available only intermittently, individuals may cope by overeating during periods of accessibility, which can contribute to overweight and obesity [70].

Food security and nutrition security are interdependent; one cannot be achieved without the other. Nutrition security is a crucial component of food security, as adequate nutrition requires more than just sufficient caloric intake – it necessitates a diverse range of macro and micronutrients to support good health and disease prevention. Recognizing this essential connection, experts increasingly use the term "*food and nutrition security*" to highlight the integration of both food availability and nutritional well-being in addressing population health needs [71].

5. Conclusion

Crop wild relatives (CWR) are invaluable genetic resources for improving wheat resilience against both biotic and abiotic stresses. They possess diverse traits that can enhance resistance to diseases, pests, drought, salinity, heat, and other environmental challenges. With climate change and the emergence of new pathogens posing significant threats to global wheat production, harnessing the genetic diversity of CWR has become increasingly important for ensuring food security and sustainable agriculture.

Through conventional breeding, wide hybridization, and molecular techniques such as marker-assisted selection and gene editing, key resistance genes from CWR have been successfully introgressed into cultivated wheat varieties. Notable successes include the incorporation of genes for resistance to rusts, powdery mildew, and wheat blast, as well as tolerance to extreme climatic conditions. These advancements not only enhance yield stability but also reduce reliance on chemical inputs, promoting eco-friendly agricultural practices.

However, despite their potential, the utilization of CWR in wheat breeding faces several challenges, including difficulties in crossbreeding, linkage drag, and the loss of genetic diversity due to habitat destruction. Therefore, conservation efforts, both in situ (natural habitats) and ex situ (gene banks), are critical for preserving these genetic resources for future breeding programs.

In conclusion, the integration of CWR into wheat breeding programs is essential for developing climate-resilient, high-yielding, and disease-resistant wheat varieties. Continued research, investment in genetic conservation, and advanced breeding strategies will be key to unlocking the full potential of CWR in addressing global food security challenges.

Acknowledgments

All authors contribute equally.

Conflict of interest

The authors declare no conflict of interest.

Notes/thanks/other declarations

The authors thank the eminent reviewers.

Author details

Anand Kumar[1*], Laxmidas Verma[2], Ravindra Kumar[3] and Sagar[1]

1 Faculty of Agricultural Sciences, GLA University, Mathura, India

2 Department of Genetics and Plant Breeding, Acharya Narendra Dev University of Agriculture and Technology, Kumarganj, India

3 Department of Agriculture, Mata Gujri College, Fatehgarh Sahib, India

*Address all correspondence to: anandkumar002014@gmail.com

References

[1] Khan M, Zhang Y, Khan S. Echoes of Survival: Climate change impact & typologies of adaptation among vulnerable communities toward climate-induced food insecurity in Pakistan. Research on World Agricultural Economy. 2025;**6**:290–318

[2] Muhammad M, Wahab A, Waheed A, Hakeem KR, Mohamed HI, Basit A, Toor MD, Liu YH, Li L, Li WJ. Navigating climate change: Exploring the dynamics between plant–soil microbiomes and their impact on plant growth and productivity. Global Change Biology. 2025;**31**:01–22

[3] Raza A, Razzaq A, Mehmood SS, Zou X, Zhang X, Lv Y, Xu J. Impact of climate change on crops adaptation and strategies to tackle its outcome: A review. Plants. 2019;**8**(34):01–29

[4] Bohra A, Kilian B, Sivasankar S, Caccamo M, Mba C, McCouch SR, Varshney RK. Reap the crop wild relatives for breeding future crops. Trends in Biotechnology. 2022;**40**:412–431

[5] Nair KP. Utilizing crop wild relatives to combat global warming. Advances in Agronomy. 2019;**153**: 175–258

[6] Shyamala G, Gobinath R, Hemalatha B, Akkalla D, Ezhil SS, Sathya V. Global research trends in environmental adaptation techniques focusing on climate change through scientometric lens. Discover Sustainability. 2025;**6**:1–19

[7] Farooq MS, Uzair M, Raza A, Habib M, Xu Y, Yousuf M, Yang SH, Ramzan Khan M. Uncovering the research gaps to alleviate the negative impacts of climate change on food security: A review. Frontiers in Plant Science. 2022;**13**:01–39

[8] Muluneh MG. Impact of climate change on biodiversity and food security: A global perspective—a review article. Agriculture & Food Security. 2021;**10**:1–25

[9] Hafeez A, Batool R, Arshad A, Khan MN, Ali S, Singh N, Garhwal V, Javed MA, Fatima E, Suleman F, Ali B. Soil and water conservation under changing climate. Challenges and Solutions of Climate Impact on Agriculture. 2025;**13**:307–328

[10] Maetens W, Poesen J, Vanmaercke M. How effective are soil conservation techniques in reducing plot runoff and soil loss in Europe and the Mediterranean? Earth-Science Reviews. 2012;**115**:21–36

[11] Sabina R, Paul J, Sharma S, Hussain N. Synthetic nitrogen fertilizer pollution: Global concerns and sustainable mitigating approaches. Agricultural Nutrient Pollution and Climate Change: Challenges and Opportunities. 2025;**11**:57–101

[12] Magadza CH. Climate change: Some likely multiple impacts in Southern Africa. Climate Change and World Food Security. 1996;**37**:449–483

[13] Bréchet LM, Salomón RL, Machacova K, Stahl C, Burban B, Goret J-Y, Steppe K, Bonal D, Janssens IA. Insights into the subdaily variations in methane, nitrous oxide and carbon dioxide fluxes from upland tropical tree stems. New Phytologist. 2025;**245**:2451–2466

[14] Ali AM. Environmental pollution and climate change implications of agricultural fertilizer use. Agricultural Nutrient Pollution and Climate Change: Challenges and Opportunities. 2025;**11**:1–28

[15] Sarika, Anand A, Meena R, Mina U, Shukla A, Sharma A. Adoption of the green energy technology for the mitigation of greenhouse gas emission: Embracing the goals of the Paris agreement. Nanomaterials and Nanoliquids: Applications in Energy and Environment. 2023;**16**:47–72

[16] Jatav HS, Minkina T, Singh SK, Singh B, Rajput VD, editors. Environmental Nexus for Resource Management, Vol. **12**. Boca Raton: CRC Press. 2024. 01–416

[17] Aziz A. Climate change: A growing challenge for food security in Pakistan. Social Science Review Archives. 2025;**3**:1390–1402

[18] Telch A. Unravelling the Nexus between Food Systems and Climate Change: A Legal Analisys. A Plea for Smart Agriculture, A" New" Organic Agriculture and A Wiser Use of Biotechnologies in the Name of Human Rights Protection. 2019

[19] Singh G, Sharma S. Desert ecology and functional aspects of desert ecosystems. Textbook of Forest Science. 2025;**19**:227–251

[20] Friedemann G, Leshem Y, Kerem L, Shacham B, Bar-Massada A, McClain KM, Bohrer G, Izhaki I. Multidimensional differentiation in foraging resource use during breeding of two sympatric top predators. Scientific Reports. 2016;**6**:01–13

[21] Dunne P. The Courage of Birds: And the Often Surprising Ways They Survive Winter. London, England: Chelsea Green Publishing; 2024

[22] Root TL, Price JT, Hall KR, Schneider SH, Rosenzweig C, Pounds JA. Fingerprints of global warming on wild animals and plants. Nature. 2003;**421**:57–60

[23] Nevo A. Measuring market power in the ready-to-eat cereal industry. Econometrica. 2001;**69**:307–342

[24] Intergovernmental Panel on Climate Change. Key dates in the AR5 schedule. Available. 2013

[25] Mercer KL, Perales HR. Evolutionary response of landraces to climate change in centers of crop diversity. Evolutionary Applications. 2010;**3**:480–493

[26] Thornton PK, Kristjanson P, Förch W, Barahona C, Cramer L, Pradhan S. Is agricultural adaptation to global change in lower-income countries on track to meet the future food production challenge? Global Environmental Change. 2018;**52**:37–48

[27] Howe JP. Behind the Curve: Science and the Politics of Global Warming. United States: University of Washington Press; 2014

[28] Zohary D, Hopf M, Weiss E. Domestication of Plants in the Old World: The Origin and Spread of Domesticated Plants in Southwest Asia, Europe, and the Mediterranean Basin. Oxford, England: Oxford University Press; 2012

[29] Yeomans L. Zooarchaeology of managed, captive, tame, and domestic birds: Shifts in human–avian

relationships. Journal of Archaeological Research. 2025;**33**:1–42

[30] He Y. Toward a Relational Foodscape in Chinese and Japanese North American Literature. Sale, England: Transnational Press London; 2025

[31] Peters Haugrud AR, Achilli AL, Martínez-Peña R, Klymiuk V. Future of durum wheat research and breeding: Insights from early career researchers. The Plant Genome. 2025;**18**:01–25

[32] Zair W Conservation of Crop Wild Relatives' Diversity in the Fertile Crescent (Doctoral dissertation, University of Birmingham). 2020

[33] Giles RJ. The Origins of Triticum Aestivum L. England: The University of Manchester (United Kingdom); 2004

[34] Turcotte MM, Araki H, Karp DS, Poveda K, Whitehead SR. The eco-evolutionary impacts of domestication and agricultural practices on wild species. Philosophical Transactions of the Royal Society B: Biological Sciences. 2017;**372**:01–09

[35] Tirnaz S, Zandberg J, Thomas WJ, Marsh J, Edwards D, Batley J. Application of crop wild relatives in modern breeding: An overview of resources, experimental and computational methodologies. Frontiers in Plant Science. 2022;**13**:01–17

[36] Endresen DTF Utilization of Plant Genetic Resources: a Lifeboat to the Gene Pool (P. 3). PhD Thesis, Department of Agriculture and Ecology, Faculty of Life Sciences, Copenhagen University, Denmark. 2011

[37] Kishii M. An update of recent use of Aegilops species in wheat breeding. Frontiers in Plant Science. 2019;**10**:585

[38] Cotter J. Troubled Harvest: Agronomy and Revolution in Mexico, *1880-2002*. Bloomsbury Publishing USA; 2003;**22**

[39] Kaur S, Jindal S, Kaur M, Chhuneja P. Utilization of wild species for wheat improvement using genomic approaches. In: Biotechnologies of crop improvement, volume 3. Genomic Approaches. Cham: Springer International Publishing; 2018. p. 105–150

[40] Saville RJ Understanding DELLA in Wheat Linking Genotype to Phenotype (Doctoral dissertation, University of East Anglia). 2011

[41] Jovovic Z, Andjelkovic V, Przulj N, Mandic D. Untapped genetic diversity of wild relatives for crop improvement. Rediscovery of Genetic and Genomic Resources for Future Food Security. 2020;**13**:25–65

[42] Ning S, Wang X, Ma C, Liao J, Ye F, Li S, Fan C, Cui Q, Wang X, Song D, Hao M. Development and characterization of small translocations between the Pm13 introgression from Aegilops longissima and wheat homoeologous A, B and D chromosomes. BMC Plant Biology. 2025;**25**:179

[43] Pramanik S, Debnath S. Development of rural landraces through mutation breeding approaches. Plant Mutagenesis: Sustainable Agriculture and Rural Landscapes. 2024;**8**:169–186

[44] Choudhary V, Kumar KK, Sahu BHIMESHWARI, Muthappa SK, Das ADA. Agronomic innovations in biotic stress management and its

combined effect with abiotic stresses in crop production. Indian Journal of Agronomy. 2021;**66**:237–257

[45] Afzal F, Chaudhari SK, Gul A, Farooq A, Ali H, Nisar S, Sarfraz B, Shehzadi KJ, Mujeeb-Kazi A. Bread wheat (Triticum aestivum L.) under biotic and abiotic stresses: An overview. Crop Production and Global Environmental Issues **13**. 2015;293–317

[46] Knott DR. The Wheat Rusts—breeding for Resistance, Vol.**12**. Berlin and Heidelberg, Germany: Springer Science & Business Media; 2012

[47] Leonard KJ, Szabo LJ. Stem rust of small grains and grasses caused by Puccinia graminis. Molecular Plant Pathology. 2005;**6**:99–111

[48] Zadoks TC. Epidemiology of wheat rust in Europe. International Journal of Pest Management B. 1967;**13**:29–46

[49] Oshunsanya SO, Nwosu NJ, Li Y. Abiotic stress in agricultural crops under climatic conditions. Sustainable Agriculture, Forest and Environmental Management. 2019;**9**;71–100

[50] Akter N, Rafiqul Islam M. Heat stress effects and management in wheat. A review. Agronomy for Sustainable Development. 2017;**37**:1–17

[51] Wahid A, Gelani S, Ashraf M, Foolad MR. Heat tolerance in plants: An overview. Environmental and Experimental Botany. 2007;**61**:199–223

[52] Ahmad Z, Waraich EA, Akhtar S, Anjum S, Ahmad T, Mahboob W, Hafeez OBA, Tapera T, Labuschagne M, Rizwan M. Physiological responses of wheat to drought stress and its mitigation approaches. Acta Physiologiae Plantarum. 2018;**40**:1–13

[53] Dhakal A, Adhikari C, Manandhar D, Bhattarai S, Shrestha S. Effect of abiotic stress in wheat: A review. Reviews in Food and Agriculture. 2021;**2**:69–72

[54] Hossain A, Skalicky M, Brestic M, Maitra S, Ashraful Alam M, Syed MA, Hossain J, Sarkar S, Saha S, Bhadra P, Shankar T. Consequences and mitigation strategies of abiotic stresses in wheat (Triticum aestivum L.) under the changing climate. Agronomy. 2021;**11**:241.

[55] Yunus MH, Firdaus A, Khan Z, Ansari MYK. Genomics-assisted breeding (GAB) for trait improvement: Unveiling genomic strategies for accelerated crop enhancement. Plant Breeding Technology: Future Trends and Challenges. 2025;**64**:138

[56] Abdallah NA, Elsharawy H, Abulela HA, Thilmony R, Abdelhadi AA, Elarabi NI. Multiplex CRISPR/Cas9-mediated genome editing to address drought tolerance in wheat. GM Crops & Food. 2025;**16**:1–17

[57] Kaya C. Plant metabolic engineering for enhanced nutrition and food security during climate uncertainty. Food and Energy Security. 2025;**14**:01–21

[58] Mujeeb-Kazi A, Kazi AG, Dundas I, Rasheed A, Ogbonnaya F, Kishii M, Bonnett D, Wang RRC, Xu S, Chen P, Mahmood T. Genetic diversity for wheat improvement as a conduit to food security. Advances in Agronomy. 2013;**122**:179–257

[59] Renzi JP, Coyne CJ, Berger J, von Wettberg E, Nelson M, Ureta S, Hernández F, Smýkal P, Brus J. How could the use of crop wild relatives in breeding increase the adaptation of

crops to marginal environments. Frontiers in Plant Science. 2022;**13**:01–21

[60] Engels JM, Thormann I. Main challenges and actions needed to improve conservation and sustainable use of our crop wild relatives. Plants. 2020;**9**:968

[61] Haridass K Exploring the Abiotic Resistance Traits of the Local Wild Wheat Relative, Thinopyrum Distichum (Doctoral dissertation, Stellenbosch University). 2024

[62] Neupane B Evaluation of Winter Wheat and Tritipyrum Germplasm to Identify Useful Variation for Biotic and Abiotic Stress Tolerance (Master's thesis, North Dakota State University). 2023

[63] Ashraf M, Munns R. Evolution of approaches to increase the salt tolerance of crops. Critical Reviews in Plant Sciences. 2022;**41**:128–160

[64] Pour-Aboughadareh A, Kianersi F, Poczai P, Moradkhani H. Potential of wild relatives of wheat: Ideal genetic resources for future breeding programs. Agronomy. 2021;**11**:1656

[65] Rouichi S, Ghanem ME, Amri M. In-situ and ex-situ conservation priorities and distribution of lentil wild relatives under climate change: A modelling approach. Journal of Applied Ecology. 2025;**62**:414–428

[66] Yao Y, Guo W, Gou J, Hu Z, Liu J, Ma J, Zong Y, Xin M, Chen W, Li Q, Wang Z. Wheat2035: Integrating pan-omics and advanced biotechnology for future wheat design. Molecular Plant. 2025;**18**:272–297

[67] Saint Ville A, Po JYT, Sen A, Bui A, Melgar-Quiñonez H. Food security and the Food Insecurity Experience Scale (FIES): Ensuring progress by 2030. Food Security. 2019;**1**:483–491

[68] Curtis T, Halford NG. Food security: The challenge of increasing wheat yield and the importance of not compromising food safety. Annals of Applied Biology. 2014;**164**:354–372

[69] Bhattacharya J, Currie J, Haider S. Poverty, food insecurity, and nutritional outcomes in children and adults. Journal of Health Economics. 2004;**23**:839–862

[70] Timmer CP. Food Security and Scarcity: Why Ending Hunger Is so Hard. Philadelphia: University of Pennsylvania Press; 2015

[71] Pérez-Escamilla R. Food and nutrition security definitions, constructs, frameworks, measurements, and applications: Global lessons. Frontiers in Public Health. 2024;**12**:01–10